RENEWALS 458-4574

DATE DUE

APR 1 6			
GAYLORD			PRINTED IN U.S.A.

The Atheist and the Holy City

The Atheist and the Holy City
Encounters and Reflections

George Klein

translated by Theodore and Ingrid Friedmann
foreword by Lewis Thomas

The MIT Press
Cambridge, Massachusetts
London, England

First MIT Press edition
English translation © 1990 Massachusetts Institute of Technology

© 1987 George Klein. First published, in Swedish, as *Ateisten och den heliga staden: Möten och tankar* (Stockholm: Albert Bonniers Förlag AB).

This book was set in New Baskerville by The MIT Press from computer disks provided by the translators, and was printed and bound in the United States of America.

Library of Congress Cataloging-in-Publication Data

Klein, George, 1925–
    [Ateisten och den heliga staden. English]
    The atheist and the holy city: encounters and reflections /
    George Klein.
        p. cm.
    Translation of: Ateisten och den heliga staden.
    ISBN 0-262-11155-1
    1. Klein, George, 1925–. 2. Oncologists—Biography. 3. Cancer.
4. Science. 5. Philosophy. I. Title
    [DNLM: 1. Medical Oncology—essays. 2. Philosophy, Medical—
essays. 3. Science—essays. W 61 K64a]
    RC265.K5413   1990
    501—dc20
    DNLM/DLC
    for Library of Congress   90-5904
                                                           CIP

for Eva

Contents

14

Peter Noll's Awareness of Death and Wisdom of Life   *189*

15

The Atheist and the Holy City   *199*

Foreword

It is one of the conventional functions of a foreword to introduce the author of the book to the reader—especially when the author is a foreigner. However, to many of the people who will want to read this book and will surely be fascinated by it, George Klein needs no introduction. The English-reading scientific community is a worldwide settlement, and George Klein is, in this millionfold company, no stranger, certainly no foreigner.

But it may be useful to say a few things to literate, non-scientist, general readers who may be curious about the phenomenon of science itself and more curious still about how the mind of an extraordinary researcher works on and off the job.

First off, George Klein ranks, within the large and steadily expanding population of biomedical scientists, as one of the best-known and most distinguished members, with discoveries and ideas to his credit that have become part of the normal curriculum for graduate students in biology and medical students everywhere. Klein's contributions to the now burgeoning disciplines or immunology, virology, cell biology, and cancer research have placed him, long since, in the position of a world-class scientist.

But more than this, it is impossible to regard as a foreigner anyone with such credentials for planetary citizenship, and on more general grounds than science alone. As this book demonstrates, George Klein cannot really be lodged in the mind as a particular national, nor even as a representative of any singular ethnic culture. He is, to put it simply, a citizen of everyplace. His colleagues and students have come to his Karolinska laboratory from every continent, and he has himself arrived at one time or another as a visiting investigator at nearly every major research center on the face of the earth.

For those readers in need of details, George Klein was born in 1925 in Hungary, grew up in a Jewish family active in the intellectual life of Budapest, received part of his education at the University of Budapest, left Hungary in 1947 for Sweden, received his M.D. degree at the Karolinska Institute in 1951, and has been engaged in research at that institution ever since, with many side trips to centers in other lands (notably ten years as a visiting professor in Israel). He has published around 800 scientific papers, and he is a member of the Royal Swedish Academy of Sciences and several foreign academies, including the National Academy of Sciences of the United States and the American Philosophical Society. The array of international awards and prizes he has received for his research is too long for these pages. He has been a member of the Nobel Assembly of the Karolinska Institute since 1960.

This book tells many things of deep interest about the working life of a major scientist, but it is considerably more than a book about science. It tells, with candor and insight, about the life going on within the private and singular mind of George Klein, a late-twentieth-century man of the world.

Lewis Thomas, M.D.

Preface

This book grew out of a foreword that I wrote for Peter Noll's book *In the Face of Death.*[1] The imminence of his death had brought him a renewed wisdom concerning life itself.

The words of this book that gush forth in his wake touch on some related and some unrelated subjects. They deal with cancer and the ways in which its study may elucidate normal cell functions, and with the many examples of peaceful coexistence between viruses and man. They take a look at that vast, astonishing river that is biology as it follows its wonderful and awesome course, inexorably and indifferently. They tell of scientists who struggle to navigate between cliffs of exhilaration and depression, between wisdom and folly.

Ganesha, the elephant-headed Hindu god of barriers, obstacles, and inhibitions, he who must be worshiped and pacified before each voyage, has been merciful to me. The words have welled up in me. I needed only step aside and watch them flow. Am I proud of them, am I ashamed of them? Both, I would say.

You words of my new language, stuttering words that I have only learned as an adult—where are you leading me? Do you want to play Cassandra? Are you driven by the vain desire to make way for better and more truthful words? I, as your witness and father, am ambivalent. My hand has often been poised to throw you into the wastepaper basket. But now I want to thank my friends who saved you from that fate, harsh critics and gentle commentators alike: Per Ahlmark, Tore Browaldh, Peter

Citron, Edith and Lars Ernster, Karl-Erik Fichtelius, Rolf Luft, Kenneth Nilsson, Benno Müller-Hill, Anders Nässil, Dagmar and Peter Reichard, and Lars Ulvenstam.

George Klein
Stockholm, January 1987

The Wisdom and Folly of Scientists

1

The Emperor's New Clothes

Mankind's research into the determinants of its own heredity is often subjected to taboos or driven by political doctrines. In recent times, there has been no shortage of voices wanting to halt this research for fear of misinterpretation and political abuse. Experience shows that man's curiosity about himself cannot be suppressed for long, but it cannot be denied that there are special perils in this field. Totalitarian dictatorships were responsible for the most frightful examples, but not even in the democracies has it been entirely spared. Every epoch and every geographical and sociological niche has had its dominant powers and pervasive concepts, which the majority hasn't hesitated to follow automatically and without question.

Those of you who would prefer to believe that these distortions have little or no effect on the pure worlds of laboratory scientists, microscopes, cell cultures, and biochemistry are in for a surprise!

Each biological species has a fixed number of chromosomes, the carriers of the hereditary factors, and every individual of the same species has the same number of chromosomes. The normal differences between the sex chromosomes and some congenital chromosomal anomalies are the only exceptions to this rule.

The study of chromosomes developed with incredible speed during the first half of the twentieth century. It was stimulated by the discovery of Mendel's laws of inheritance and by the development of the modern microscope. The identification of chromosomes as the carriers of the Mendelian hereditary factors—the genes—and the subsequent integration of cross-breeding experiments with microscopic observations are among the most beautiful chapters in the history of the sciences. Bean plants and fruit flies were among the favorite objects of early genetic studies. Mammals made their entrance quite late, led by the smallest and most rapidly breeding among them: the mouse.

Certain species earned bad reputations from the very beginning as rather difficult to study. It was cumbersome to make good preparations of their cells. The chromosomes clumped together, and one generally could see only ugly dots instead of the beautiful, slender, rod-shaped individual chromosomes. Cell geneticists were driven to despair. A few animal species are even now still quite inaccessible—the chicken is an example.

As recently as the early 1950s, the chromosomes of man were considered to be among the most difficult, or were even thought to be completely hopeless. Despite that, people tried to study them. Techniques improved slowly, and after a while it became possible to count and characterize the human chromosomes, at least grossly. It was concluded that humans have 48 chromosomes. This number was confirmed by many published studies, quoted in lectures, written in textbooks, and considered to be a well-established fact.

In the early 1950s, a distinguished Swedish chromosome researcher in Lund, Albert Levan, became interested in mammalian chromosomes. He had previously worked with plant chromosomes, in the best traditions of Lund. During a trip to the United States, he came into contact with the mouse geneticist Theodore Hauschka at the Institute for Cancer Research in Philadelphia, where I myself spent a few months during my doctoral studies. Hauschka was one of my closest contacts. My task there was to transform solid mouse tumors into a so-called suspension culture in which the cells could multiply freely in the fluid of the abdominal cavity, almost like a culture of microorganisms. Hauschka became very enthusiastic when he saw how easily these cells could be grown, harvested, and manipulated. After Levan arrived, they used these mouse tumor cells to make some of the most beautiful preparations of mammalian chromosomes ever seen.

Their work opened the way to a new area of research. After the first thrilling experiences with the "suspension cell" tumors, scientists became more daring and extended their work to solid tumors and gradually even to normal mouse tissues. But the human chromosomes remained inaccessible.

After his return home to Sweden, Levan began working with a guest scientist, J. H. Tjio, who had been born in Indonesia to Chinese parents, raised in Holland, and employed in a permanent position in Spain, and who was married to a woman from Iceland. Tjio had no deep roots in any country. His ties were, if anything, rather negative. His hypersensitivity to real or presumed racial discrimination was legendary. He spied racists behind every bush, even in democratic Sweden. Even so, he was happier there, especially in Levan's laboratory, than anywhere else. It was a place where he could devote himself to his special art. He succeeded in making chromosome preparations of a quality previously unheard of, and he photographed them in a way that astonished even the most experienced chromosome researchers. Tjio and Levan together took on the taboo human cells, and, amazingly enough, found them not so difficult. Had all the previous investigators been hampered by their own preconceived notions? Was it because they hadn't wanted, or been able, to undertake the exhaustive and hard work necessary to make good chromosome preparations from such a new biological source, or had they simply not had enough strength to endure all the inevitable setbacks that every new field of research faces at the beginning?

Tjio and Levan's preparations of human chromosomes were of such good quality that every single one of the isolated chromosomes could be cut out of a photograph, arranged next to all the others like the pipes of an organ, and counted. Tjio and Levan counted the chromosomes, and amazingly there were 46, not 48! But how could all those previous publications have described 48? Tjio and Levan reviewed the earlier literature. In the most popular text, Cyril Darlington's *Cytology*, there was a picture of chromosomes reproduced from a paper by the eminent chromosome researcher T. C. Hsu. The figure legend stated that the photo presented a "typical complement of 48 human chromosomes." Tjio and Levan counted the chromosomes in the picture. There were 46—no more, no less!

The incorrect count had been accepted by the referees and the editors of one of the most respected journals of the time. The same mistake was repeated when the picture was later reproduced in the best textbook on the subject. Neither authors nor readers questioned it until the truth burst forth in the form of a five-page article entitled "The Human Chromosome Number," published by Tjio and Levan in the journal *Hereditas*.[1] That article was the opening shot in the far-reaching modern study of human chromosomes, which was to become so productive in the ensuing years. It was the first demonstration that studying human chromosomes was not only possible but extraordinarily appropriate. Today the human has reached full equality with the bean plant and the fruit fly as one of the most useful objects for chromosomal studies!

Tjio and Levan received many letters after their sensational publication. Chromosome researchers wrote from all over the world. Many said that they, too, had counted 46 chromosomes, but being fully convinced that they must be wrong they had recounted over and over until the number had become 48. One Japanese scientist was so distressed by the risk of losing face that he continued to publish papers in which he claimed that the number of chromosomes varied in different people, and that there were people with 46, 48, and even 47 chromosomes. This, of course, has nothing to do with the truth. All humans have 46 chromosomes—no more, no less. The only deviations from this number are found in association with pathological conditions, most of which are very severe and lead to death early in fetal life.

How often do otherwise critical scientists read material dealing with commonly accepted postulates without examining the underlying facts? This is a troubling question. Some years ago, *Science* published the results of a sociological study in which scientists had been asked to comment on a group of references. In half the cases, the readers had been told the origin of the citations; in half, they had not. The study reached the following four conclusions:

1. If one has a high regard for an author, one is inclined to agree with him *no matter what he says*.

2. If one agrees with a conclusion, one is inclined to have a high regard for the author *no matter who he is.*

3. If one has little regard for an author, one will not readily agree with him *no matter what he says.*

4. If one doesn't agree with a statement, one easily develops a low regard for the author *no matter who he is.*

These points were perhaps self-evident from the very beginning, even before this extensive and well-documented research study was carried out. On the whole, one often doesn't even bother reading about something that is taken for granted. Whether or not one agrees with the author of a new study, one assumes that "the matter is settled."

One of our century's greatest biologists, Francis Crick, who together with James Watson discovered the double-helix structure of DNA, loves to play little practical jokes. This has been somewhat of a tradition among the "fathers" of molecular biology. They have enjoyed mixing seriousness with levity and rigorous thinking with sarcasm or even self-irony. This has served as a source of inspiration to themselves and to others. Long after Crick and his close friend and colleague Sidney Brenner had proved that the genetic code consisted of three basic letters, and even after research in the field had advanced a great deal further, some of the basic experiments still remained unpublished. It was obvious that they ought to be published sooner or later, at least for historical reasons. When Brenner and Crick somewhat reluctantly wrote up their experiments for publication, the result was a manuscript of some 80 pages. What journal would possibly think of accepting such a mass of material that merely provided further proof of a concept that was already commonly accepted? Crick decided to avail himself of his privileges as a member of the Royal Society (the British academy of sciences) and publish the article in its *Proceedings*. But both he and Brenner doubted that anyone would read this article. They decided to do a little experiment. In the middle of a long section full of many rather tedious details, they inserted a small reference in parentheses. Instead of an ordinary journal reference, they cited "Leonardo da Vinci, personal communication." They then submitted the article to the journal. It was immediately accepted. The "references" were approved by the journal's scientific reviewers and by the editor-in-chief without comment. The Leonardo citation was detected only at the last minute by a technical assistant whose job it was to check all references for spelling errors before material was sent to the printer.

There is an institute of behavioral studies at Yale University that has been interested in conformity as a sociological phenomenon. In one famous experiment, volunteer subjects were brought in, one at a time. Each subject was introduced into the same group of nineteen people and was led to believe that the group consisted of subjects chosen at random, when in fact the nineteen others were really lab personnel. They sat in a classroom. Someone impersonating a teacher drew two parallel lines, differing only very slightly in length, on the blackboard, and asked the group which line was longer. It had been arranged that the subject of the experiment always was the last to answer. The nineteen who answered first unanimously answered incorrectly. The study was designed to determine how often the volunteer "voted" with the group and how often he used his powers of observation. In the beginning, the subject always went along with the group. The study continued with similar questions, but with increasingly greater differences between reality and pretense. The results demonstrated an overwhelming degree of "conformism." Virtually all subjects went along with the group to a great extent, regardless of their social background or level of education. They finally "awoke" rather late, when the discrepancies between pretense and reality had grown grotesque. The tale of the emperor's new clothes is therefore, to a great extent, valid even in democratic societies which believe that they have long since rid themselves of "imperial authority."

In authoritarian societies, these kinds of psychological mechanisms have led to the construction of imaginary "castles in the air" that are not brought to ruin until a very late stage and only after a great deal of damage has been done. An honored biologist and biochemist in prewar Germany, one Dr. Moewus, published a lengthy and famous series of articles in which he claimed to have isolated and purified a whole new class of previously unknown substances that he claimed were secreted by protozoa. These substances were said to function as signals for sexual reproduction and for other aspects of the life cycle. He gave them complicated new names and incorporated them into an intricate scheme in which their structure and their biological activities were mapped in great detail. A new branch of science had apparently been developed, complete with its own highly qualified groups of scientists and students. Moewus became regarded internationally as one of Germany's greatest biochemists.

It was not until after the war that American scientists tried to reproduce Moewus' results. Not even the simplest basic experiments worked! Other scientists went to Moewus' laboratory and made tremendous efforts to repeat his studies. They really wanted very much to verify them, since Moewus' work seemed to combine aesthetic beauty, biochemical precision, and biological plausibility. Only nature itself could have devised such a multifaceted system, certainly not the human brain! Then, at a symposium in Wales organized by the English Society of Experimental Biology during the latter half of the 1950s, I heard two Americans give a memorial speech to Moewus' theory. They proposed that a monument be erected and inscribed RIP, and that it should be written underneath that never before in the history of science had such an enormous effort been made to confirm such a beautiful and appealing theory that eventually was shown to be built on such an utterly vacuous foundation.

The authority that Moewus and the officials of his institution wielded within the German academic system made skepticism about his work virtually impossible, not only on political and social but even on psychological grounds. It just didn't seem imaginable that a scientist of such stature could construct what seemed to be a whole new branch of science out of such complete fabrications. Young researchers who couldn't reproduce the results must therefore have thought that *they* were the ones who were wrong—very much like those who counted chromosomes again and again until they got the presumed "correct" but actually wrong number.

But what can happen in less authoritarian academic systems? Do they run similar risks? Of course they do, as the episode with chromosome counting has already shown; but not to the same degree. In the relatively democratic American world of science, a young scientist can always gain acceptance by tearing down an established authority, as long as he can prove that he's really correct. This can, in its own right, have just as valuable and constructive an effect as a more positive achievement might have. On the other hand, it can be quite difficult, especially for relatively unknown scientists, to claim results and interpretations that contradict generally accepted notions of the times, even in that more liberal American system.

Today every schoolchild knows that the language of genetics, the genetic code, is written in the letters of DNA molecules. But this fact wasn't at all obvious at the beginning. It wasn't until a decade after the first definitive experiments were published that such information "became knowledge," as Gunther Stent expressed it. During the early 1940s, when the discovery was first made, the nucleic acids DNA and RNA had already been "definitively ruled out" as likely carriers of genetic information!

It is easy to understand the reasons for that. The most important organic components of all living organisms are proteins. They consist of twenty building blocks, the amino acids—an alphabet of twenty letters. But the nucleic acids have only four component letters. How can an alphabet be built of four letters? Furthermore, the acknowledged experts in nucleic-acid chemistry several decades earlier, Levene and Bass, had declared the nucleic acids to be uninteresting molecules, and their opinion had set the tone. In their view, DNA and RNA had a monotonous and repetitive structure similar to that of the ground substance of connective tissues and bony structures. How could the richness and variety of the whole living universe be contained in something so monotonous?

The definitive experiment was published in 1944 by three scientists, Avery, McLeod, and McCarty, who were working at the Rockefeller Institute. They based their experiments on an earlier study carried out in the 1920s by Frederick Griffith, an Englishman who had shown that pneumococci of one certain type could be converted to another if they were treated with heated cell extracts from the second bacteria. The change was stable and heritable, and the phenomenon was called "bacterial transformation." The great accomplishment of the Avery group was to show that the transforming substance was DNA and not, as had been thought, protein.

The scientific world had taken notice of Avery's experiment but didn't fully understand its significance. At first, very few of the scientists who were active at the time and who later were to become some of the most enthusiastic spokesmen for the new DNA-based genetics realized the importance of this genetic transformation. They simply weren't ready just yet to have a fresh look at this "boring" molecule of DNA. It was much easier to believe that Avery's preparation was impure—DNA was known to be difficult to purify. How could one be sure that the preparation didn't contain some molecules of an important protein? Avery had indeed shown that an enzyme capable of degrading DNA also destroyed the transforming activity but that another enzyme that broke down protein had no such effect. But wasn't it possible that DNA's role was to protect an important protein that actually carried the genetic information? Couldn't the protected protein have become resistant to protein-degrading enzymes? If that were the case, the breakdown of the protecting DNA could destroy the transforming effect. And who, after all, could know if the enzymes that Avery and his colleagues had used were sufficiently pure to have only their expected effect?

Avery died before the great reevaluation came. The protective walls around the earlier concept came tumbling down in the 1950s when it became possible to separate viruses into their nucleic-acid and protein components. It was found that virus infections could be transmitted to new cells through viral nucleic acid but not through viral protein. Even that finding didn't silence the critics immediately, but additional proof relevant to so many animal and plant viruses streamed in so quickly that the supporters of "protein genetics" finally had to give up. Nevertheless, a certain skepticism remained in the air for several more years. Heinz Fraenkel-Conrat, one of the pioneers who showed that the tobacco-plant "mosaic virus" could infect plants by means of its purified nucleic acid, became the victim of a classical practical joke by his own students. To test their chief's confidence, they sent him a telegram just before a lecture: "Be very careful what you say! The last experiment suggests that virus protein may be infectious!" Did that make Fraenkel-Conrat more cautious? No—he realized immediately that it was a prank. But only six months earlier, he would probably have vacillated.

Critics of the nucleic-acid theory didn't understand the principle—so simple now but so incomprehensible at the time—that the code words of genetic information do not consist of one basic letter, as was first believed because of the analogy to the phonetic alphabet, but rather of a combination of three basic letters. By means of such an alphabet, one achieves a code that is much "richer" than it needs to be. If one makes all conceivable three-way combinations of four basic letters, including repetitions, one gets 64 such combinations— far more than are needed to code for twenty amino acids plus a couple of "stop signs." In the language of information theory, the genetic code is therefore "degenerate"—in other words, more than one code word can specify a particular amino acid.

In his description of this development, Gunther Stent points out that experimental facts must always be integrated into the existing universe of ideas before information can be trans- formed into knowledge. This is, after all, merely a special case of the development of cognitive function.

On an expedition in New Guinea, in search of a new virus, Carleton Gajdusek (the discoverer of the disease kuru) lived among a truly Stone Age people for several months. As was his custom, he befriended members of the local tribes and learned their language. After some time, he brought out his Polaroid camera. When he pulled out the first picture of the assembled family, everyone stared at it without grasping its meaning. They had never before seen a picture! They had no mirrors, and the mountain streams were far too rapid for them to be able to see their reflections clearly in the water. After a lot of head- shaking, one man suddenly shouted: "But there is my pig!" No sooner had he said that than some of the others also saw the pig. There was great merriment, the boundless joy of new discovery. With the pig as the point of reference, the whole picture gradually took shape; people recognized one another, and at last, with a degree of reluctance, they even recognized themselves.

2

Confrontation through an Emissary

Erkenne die Zusammenhänge der Dinge und die Gesetze der Handlungen der Menschen, damit Du wissest, was Du tuest.

Recognize the connections of things and the laws of conduct of men, so that you may know what you are doing.

Szilard's first commandment [1]

Sometime during the latter part of the 1960s, when the Vietnam war was at its height, I received a letter from an American colleague, one Dr. B, whom I barely knew. Relatively young, he had already made a major contribution to a particular field of research, and he was known as an innovative scientist, unafraid to strike out in new directions. Our interests had developed along very different lines, and I hadn't heard from him for several years. I had been told by others that he was heavily involved in the protest movement against the war.

B asked me to look into something for him. A certain Dr. G, working at a Swedish university, had published a paper in an American military medicine journal that had upset B very much. Considered on its own it was a superb biochemical study, but it seemed to have been written by a fanatic Nazi and a warmonger. Could I please find out more about G's activities? Hadn't anything been done about his ravings? Surely a country like Sweden wouldn't tolerate something like that, would it?

I had never heard of G before, but I went to the library to look up the journal. The article was a solid technical paper that described genetically determined biochemical differences among the various human races—white, black, yellow, Indian. It was a beautiful combination of genetics and biochemistry. The author concluded that these kinds of biochemical differences could be used to develop chemical weapons which might exterminate people of one race selectively without harming others.

An old friend of mine, L, worked at G's institution. I cut off the letterhead and the signature, glued the remainder of B's letter to another piece of paper, and mailed it to L with some questions. Who was G? Why had he written the article?

L answered the next day. He was dumbfounded. G was one of his most decent colleagues, a committed democrat and a true humanist to his very fingertips. Any notions of racial discrimination were completely foreign to him. On the contrary, he always fought actively against all anti-democratic tendencies wherever he found them. L took the liberty of showing my letter to G, who was shocked. He wondered if my anonymous correspondent was even literate. The aim of the article should have been obvious to all thoughtful people. It was, in fact, to draw attention to the great danger of this kind of biochemistry. It was an alarm signal for the need to prohibit the use of scientific discoveries for racial purposes before it was too late.

I cut off the top and the bottom of L's letter, made a quick translation, and sent it to B. I didn't have to wait long for his answer: "If that's what G meant, why didn't he say so in his article?"

The cutting of letterheads and signatures was becoming a matter of routine. The letter went to L.

G answered through L: "Who am I to say something like that? I'm merely a biochemist. There must to be some among the military and political leaders of the great powers who are more qualified to take a stand on that question than I am. The purpose of publishing the article in a military medicine journal was precisely that: to get responsible people to react and to take further measures to prevent abuse."

I could only shake my head in disbelief.

After I revealed B's identity, the two gentlemen began a direct correspondence with each other, and I lost track of the whole affair. But I was still amazed. Was such naiveté really conceivable in Sweden in the 1960s? Where had G and his sources of information been hiding during the last half-century? Was it really possible for an eminent professional to be so completely ignorant of the historical and political course of events?

But in later years I came to realize that I had been unfair. If one lives in a relatively homogeneous society with a universally accepted system of norms, it is difficult to appreciate the fact that the same system of norms isn't necessarily operative in other societies or during other periods of history. The world of the scientist is no exception to this general rule. We will take a look at several less innocuous examples.

3

One Should Never Do Anything at All

Deine Taten sollen gericht sein auf ein würdiges Ziel, doch sollst Du nicht fragen ob sie es erreichen; sie seien Vorbild und Beispiel, nicht Mittel zum Zweck.

Let your acts be directed towards a worthy goal, but do not ask if they will reach it; they are to be models and examples, not means to an end.

Szilard's second commandment [1]

I met Leo Szilard for the first time in a colleague's home while I was on my way to a meeting in the United States in 1955. He has been called "the father of the atom bomb" and "the organizer of the Manhattan Project." After Hiroshima, he left physics and devoted himself to two activities: biology and politics. He carried out his political work mostly behind the scenes, through personal contacts and private channels, just as he had done before and during his work on the Manhattan Project. But the purpose of his activity was now reversed. Szilard carried out his underground activities like a mole and devoted himself to promoting agreements on nuclear disarmament and to the organization of the Pugwash Conferences, meetings at which Soviet and American scientists had the opportunity to establish personal contacts and to discuss steps to steer their governments in peace-seeking directions. He took on the organizing of a pressure lobby in Washington, the Council for a Livable World, whose purpose was to monitor the political process and to work for peace.

We had a short conversation about immunology, a subject that had begun to interest Szilard. He said that he wanted to understand more about the relationship between tumors and the immune system, and he promised to visit me in Stockholm. We parted as if we had always known each other, but I didn't believe for one minute that he would really come to visit me.

Six months later, the door to my office in Stockholm flew open without the slightest warning and in came Szilard. He had neither telephoned nor written. He sat down across from me, as if we had arranged a meeting, took out a little notebook, and asked me to tell him all the latest facts about my field. He took some notes, made some minor comments, and asked some perfunctory questions. I introduced him to my younger colleagues. He talked to them all in the same quiet manner, even to the youngest one. Then he left, as suddenly as he had arrived. We didn't know if he had become interested or if he intended to return. The next day, the door opened again. Szilard sat down. He'd been pondering what he had heard. He tried to interpret some of our experiments, and he suggested still others. About half of what he said was based on a misunderstanding or on insufficient insight into our experimental system, but the other half was really ingenious. As we listened to him, we couldn't understand how his simple explanations had escaped us, or how we ourselves hadn't thought of the experiments that he was suggesting—experiments that could give us direct answers to the questions we were posing.

We soon began to understand Szilard's behavior. When he got up suddenly and left, it meant only that he was tired. Sometimes he would return the following day, sometimes after several or many months. We gratefully accepted the fact that we were on his list of people to visit, and we didn't expect him to announce himself in advance. Szilard was the world's first, and possibly last, "roving biologist." His journeys took him to the Pasteur Institute and to a whole string of laboratories in the United States, Switzerland, and England. He behaved exactly the same everywhere. During the late 1950s, Szilard was mentioned in footnotes in many publications, among them some of the most important papers in biology. At times the authors thanked him for suggestions that had led to their experiments; at other times the authors acknowledged that Szilard's interpretations had clarified their results. On occasion he was even thanked for providing a practical suggestion that had led to the development of a new method.

Szilard rarely talked about his past—anyone who wanted to know could read about it in connection with the Manhattan Project. Even more rarely did he speak Hungarian, our common language. At times, however, and without any accent, a few Hungarian words slipped out—some incisive and clever, some rather low-keyed. I once asked him when he had left Hungary. It was around Christmastime in 1919. He was on his way to study in Berlin, then the mecca for physics. He was sitting by himself on the train, feeling a bit depressed. The future for Hungary certainly seemed dark after a war, a revolution, and a counter-revolution, but it was nonetheless risky for a young man to go out into the great unknown world. Across from him sat an old peasant who looked as if he had never traveled further than the next village. He noticed that the young man seemed unhappy and asked why. Szilard answered that he was about to leave Hungary, possibly forever. The old man said knowingly: "You should be pleased about it. As long as you live, you'll remember this day as the happiest one of your life." The peasant was a Hungarian emigrant who had settled in Canada and was making his first return visit to his homeland in forty years.

No, Szilard didn't have very many good things to say about his homeland, although many of the Jewish-Hungarian physicists and mathematicians in the United States belonged to his circle of closest friends—among them Eugene Wigner, Edward Teller, and John von Neumann. Two of them had attended the same high school as Szilard in Budapest. Szilard was once asked in an interview if he had had an inspiring mathematics teacher. "Oh, no, he was a complete idiot," answered Szilard. Mathematics classes were intolerably boring. The three friends, all of them talented in mathematics, therefore decided to organize their own mathematics study group. Their long journey toward the stars had its beginning there.

Nevertheless, as Szilard points out in the documents and memoirs published after his death,[1] the value system of the society in which he grew up—a liberal Jewish middle-class environment—was a strong stimulus to a young man anxious to devote himself to science. But it isn't only the roots of Szilard's scientific interests that can be traced to this cultural environment. Like other great scientists who originated in the same cultural circles, Szilard had very broad intellectual interests that he could pull together instantly into all sorts of different combinations. This led to expressions that others would come to call "Szilardisms"—remarks that could be humorous, provocative, cynical, poetic, or even withering. One might love or hate them, but it was impossible to be indifferent to them. They were usually coupled with an unbelievably acute sense of observation, especially with respect to people and their stated or concealed motives.

Possibly the most extraordinary quality of this always reserved, polite, and seemingly reticent Central European gentleman was his complete lack of shyness or humility when it came to following rationally drawn inferences to their ultimate conclusions. The Manhattan Project is only one of many examples, even though it became the most fateful. Because of that special quality, Szilard was not only willing to take on the role of the conscience of the world; he *was* the conscience of the world. In contrast to lesser consciences, he wasn't satisfied merely to preach; he also initiated and supervised more practical work directed toward his desired goal. When he realized, even while he was in Germany, that a nuclear chain reaction was possible, he immediately recognized the danger that Hitler would produce an atomic bomb. His fear eventually turned out to have been unwarranted, but only in retrospect, and that fact doesn't lessen the potential magnitude of the danger as Szilard then saw it. That Germany did not develop an atomic bomb was due, in part, to the unwillingness of the imperious German professors to cooperate with one another, but perhaps even more to the frenzied disgust of Hitler's faithful physicists for "Jewish"—in other words, nuclear—physics.

Szilard's involvement as organizer of the American atomic-bomb project was based entirely on the conviction that it was necessary to beat Hitler to the bomb if Western civilization was to survive. Szilard's success depended equally on his scientific competence and on his political skill. He had a rare ability to motivate others to devote themselves to the project, either with scientific arguments or, if necessary, with political reasoning.

This unique skill was based on, among other things, the injunction that Szilard's "first commandment" (see the opening page of chapter 2) imposes on an individual. He could easily empathize with the thoughts of others. This characteristic was quite common among Central European Jews, who had had ample opportunity to develop such a trait during their many centuries as victims of discrimination and persecution. However, in Szilard's case, that quality was combined with energy and determination to achieve the goals that he considered vital for humanity. His only limitation lay in the fact that he truly believed that scientists could not only produce an atomic bomb but also control its eventual use.

The beginnings of the Manhattan Project came when Szilard convinced the initially reluctant Einstein to write the letter (dated August 2, 1939) that led President Roosevelt to establish a committee to look into the possibility of building the bomb. Szilard's proposal to produce a chain reaction that would be useful for military purposes was first met with strong opposition from the committee's Army representatives, who considered the project senseless. Who could imagine developing a new kind of weapon, based on a principle that was still untested, while a world war was going on?

Szilard didn't give up. His goal was clear, despite the fact that he understood that it would lead to an irrevocable change in the future of the world—a change whose significance would reach far beyond the Second World War. He described the first chain reaction, produced on March 3, 1939, in the following way: "Everything was ready and all we had to do was to turn a switch, lean back, and watch the screen of a television tube. If flashes of light appeared on the screen, that would mean that neutrons were emitted in the fission process of uranium, and this in turn would mean that the large scale liberation of atomic energy was just around the corner. We turned the switch and we saw the flashes. We watched them for a little while and then we switched everything off and went home. That night there was very little doubt in my mind that the world was headed for grief."

One of Szilard's most important aphorisms was that it is not necessary to be more clever than other people, but it is important to have the right idea one day before everybody else. There had already been many examples of this in his life. In contrast to most of his colleagues, he had realized as far back as 1930, in Berlin, that Hitler would eventually rise to power. He had come to that conclusion through his observation that a vast majority of the population talked in pragmatic and utilitarian tones, unburdened by any ethical values. Rather than object, they were usually content to ask only: What good could my opposition possibly do? After thinking about the question, most decided to remain silent.

When it came to the Manhattan Project, Szilard realized clearly that entry into the atomic age would cause many serious problems. For a time he believed that the scientists could devise ways to control the development of the atomic age, even after the bomb became a reality. But he certainly understood, "one day before everyone else," that this was not going to be the case.

The Manhattan Project turned out to be an enormous technical and organizational success. By 1944, however, it was already quite obvious that Germany was going to lose the war, that there was no German atomic bomb, and that no such bomb was being developed. The American bomb wasn't yet ready, but it was close to becoming a reality. Szilard began to wonder what sense there was in continuing with the bomb project. How would the bomb be used if the first one became available before the end of the war with Japan? After it became clear that there would be no German atomic bomb, the incentive for Szilard, Einstein, and many other physicists to continue vanished. But the project proceeded at full speed. During the spring of 1945, Szilard came to the conclusion that his influence over the use of the bomb was minimal or nonexistent, and in a last desperate move he convinced Einstein to write a new letter to President Roosevelt. In that letter (dated March 25, 1945) Einstein wrote that, because of considerations of secrecy, he hadn't been able to obtain all the relevant information from Szilard, but that he nevertheless wanted to inform the president of the fear that Szilard and other scientists had concerning the increasingly inadequate communications between the scientists and representatives of the government.

He asked Roosevelt to meet with Szilard as soon as possible and to listen to his case.

Szilard tried to send the letter to the president through Mrs. Roosevelt, who had acted as an intermediary between the physicists and the president. Szilard was given an appointment with her for May 8, but just days after he received notification for his appointment, on April 12, the president died.

Szilard tried desperately to establish contacts with President Truman, but none of his Washington connections had any channels of communication to the previously obscure vice-president from Missouri. With his characteristic ingenuity, Szilard inquired if anyone connected with the Manhattan Project came from Kansas City. He made one such contact. Three days later he got an appointment to go to the White House. He didn't get as far as the president. Rather, he was referred to James Byrnes. Byrnes was to become secretary of state, bu Szilard didn't know that. He presumed that Byrnes was to become responsible for all uses of nuclear power, and that that was why he had been sent to him. Szilard tried to appeal to the political goals that he thought Byrnes supported. He emphasized that it would be a great mistake to use the bomb—it would be tantamount to disclosing a great and dangerous secret before deciding how nuclear power should be dealt with in the future. America would risk losing her position of leadership. Byrnes didn't agree at all with Szilard. He considered it important that results of the project be demonstrated right away. Congress had spent two billion dollars on the project, and if the bomb were not used the public might think that the project had been a failure. Byrnes also thought that the USSR would treat the Eastern European countries more moderately out of an increased respect for America's military strength after an atomic bomb was dropped. Szilard was, in his own words, flabbergasted at the presumption that such saber-rattling would make the Soviets "friendlier."

Szilard tried to persuade Byrnes that the use of the atomic bomb could lead to a nuclear arms race between the United States and the Soviet Union. That argument didn't meet with any greater sympathy, and he left Byrnes' office deeply depressed. He wished that he had been born in the United States and become a politician, and that Byrnes had been born in Hungary and studied physics. In that case the world wouldn't have seen any atomic bombs and the threat of a nuclear arms race would never have arisen. With the advantage of 40 years of perspective one might wonder, but it does seem likely that history would have taken a different turn.

Szilard then tried to persuade J. Robert Oppenheimer, the scientific director of the Manhattan Project, to use his influence to prevent the use of the bomb against Japanese cities, but Oppenheimer didn't agree with Szilard. Another physicist who didn't share Szilard's view that scientists should block the military use of the bomb was his countryman and old friend Edward Teller, later known as "the father of the hydrogen bomb." On July 2, 1945, one month before Hiroshima, Teller wrote that if Szilard could convince him that moral objections to the use of the bomb were correct, he wouldn't protest but instead he would immediately stop working on the project. He added: "The things we are working on are so terrible that no amount of protesting or fiddling with politics will save our souls."

But Teller remained unconvinced by Szilard's objections. He thought it impossible to eliminate any specific weapon from the arsenal. Rather, he thought that war itself would have to be eliminated if humanity was to survive. The more dangerous a weapon might be, the more likely that it would be used in an actual conflict. No pacts or agreements could prevent that. Teller therefore thought that the only hope lay in a demonstration of the extraordinary power of the atomic bomb. Only in that way could everyone come to realize that the next war would be catastrophic for all of mankind. "For this purpose," he wrote, "actual combat use might even be the best thing."

Teller also had a completely different view regarding the responsibility of scientists: "The accident that we worked out this dreadful thing should not give us the responsibility of having a voice in how it is to be used. This responsibility must in the end be shifted to the people as a whole, and that can be done only by making the facts known." Accordingly, Teller thought that the shroud of secrecy should be lifted from the atomic bomb as soon as possible.

In the final analysis, Teller said that it would be a mistake "if I tried to say how to tie the little toe of the ghost to the bottle from which we just helped it to escape."

Early in May of 1945, a committee was appointed to consider the possible use of the bomb. Secretary of Defense Henry Stimson was chairman, and Oppenheimer was one of the scientific advisors. Szilard was extremely unhappy about the composition of the committee, since he believed that several of the members were already committed to the use of the bomb. They considered it their obligation to prove that the money spent had in fact been used productively. Szilard had a great deal of respect for Stimson, but regarded his position as untenable. Stimson thought that, since the United States had only two bombs, it would be risky if the Japanese were made aware of the existence of the bombs by a demonstration under suitable conditions— that is to say, without dropping it onto a city. If the demonstration failed, everyone would conclude that the whole thing was a bluff and would scoff at the United States. However, according to Szilard, more bombs could could quickly be made, and the risk that none of them would work was nonexistent.

After all these adversities, Szilard realized that it was impossible to prevent the government from using the bomb. He therefore wrote a letter of protest and circulated it among scientists, in violation of security regulations and at great personal risk. Fifty-three physicists (including most of those in positions of leadership), and many leading biologists, signed the petition. However, the chemists refused to sign. Szilard tried to have the document delivered to Truman, despite clear warnings from the Army against such an action and hints that this might constitute a violation of his contract and a breach of security.

Truman was in Potsdam, and Szilard was assured that the petition would be delivered to him there. It probably never was, however—in any case, not before the bomb was dropped over Hiroshima. Even afterward, the Army continued to threaten

Szilard. They didn't want him to make the petition public and thereby disclose that a great many of the scientists disagreed with the Army. But again Szilard refused to obey, and the petition was made public.

The day after Hiroshima, Szilard went to see Einstein. Einstein was terribly upset and said to Szilard: "You see—the old Chinese were right. They said, 'One should never do anything at all.'"

After the war, Szilard came to believe that biology was the most important science for the future, and he therefore decided to devote half of his time to it. Although he was a physicist, he acquired great insight into biology and became one of the pioneers of modern molecular biology. The other half of his time he devoted to politics. Bitter experiences before and after Hiroshima didn't shake Szilard's conviction that the world should be, and can be, governed by reason, and that scientists are more likely to act rationally than most others. Herein lay Szilard's greatest hope and probably his greatest mistake. However, it is a fact that the lobby that he established in Washington—the Council for a Livable World—exists and is very active even today under the leadership of one of the great biochemists of our time, Matthew Meselson.

4

Szilard Plays Chess with Death

Führe das Leben mit leichter Hand und sei bereit fortzugehen, wann immer Du gerufen wirst.
Lead your life with a gentle hand and be ready to leave whenever you are called.
Szilard's tenth commandment[1]

One weekday morning in 1959, my door opened and Leo Szilard came in, unannounced as usual. He had just come from Vienna. He told me he was writing a book that he was going to call *The Voice of the Dolphins*, a utopian vision that was to include his ideas on how rational people could prevent an atomic war between the United States and the Soviet Union. He was also organizing a Pugwash Conference at which Russian and American scientists could get together to discuss the problem of how to achieve peace. He had been promised a meeting with Khrushchev to discuss the test-ban treaty, which to a great extent was influenced or even initiated by physicists at the Pugwash Conferences, with Szilard as their unfailing driving force. After a short account of his activities, he took out his little notebook and asked me about the latest news in tumor biology and immunology.

I told him what I knew. I noticed that he looked pale and tired. Even though Szilard loved to eat well, not even my suggestion to have lunch together in a first-rate restaurant seemed to cheer him up. He answered dryly: "Isn't it true that the nearest restaurant with edible food is located in Paris?"

He had to continue to Uppsala. I asked one of my students to take him to the train station. As usual, we had no idea whether or not he was coming back.

The next day, my door opened again and Szilard sat down and took out his notebook. But even before we started to talk, he wanted to know what my assistant, the one who had driven him to the train station, did in the laboratory.

I replied: "He's not exactly one of those highly articulate and intellectual types that you usually like, but he is a superb experimentalist."

"Throw him out," Szilard said.

"Why?"

"He's unintelligent."

"He is *not* unintelligent! He's very good and clever with experiments. His studies are some of the most original ever done here, much more so than those done by some of the intelligent and articulate students that you prefer."

"You're wrong. Throw him out! A stupid person like that could lower the intellectual atmosphere in your group."

We changed the subject. Szilard was, of course, completely wrong. The student in question later published one of the most original dissertations ever carried out at our institution. But Szilard couldn't at all understand his inarticulate, practical, probing, and intuitive intelligence—the kind that is so essential in experimental research.

As usual, Szilard offered some suggestions on the subjects that I had mentioned to him the day before; as always, half were brilliant and the other half truly mad. Then he asked to use the toilet. When he came back, I also went to the bathroom and noticed some traces of blood. I returned and sat down across from him.

"I'm sorry, I know that it's not my business, but I think that you're not well. I noticed it even yesterday. Now I realize that you're bleeding. What's wrong?"

"I'm bleeding from the bladder."

"How long has that been going on?"

"Six months."

"My God! Have you been to see a doctor?"

"No, not at all! What good would that do? I have no confidence in American doctors. All they want to do is to make money. They love to operate. Besides, I can't be bothered. I have more important things to do. Plans for the Pugwash

Conferences are now going ahead, and I'm hoping to have a meeting with Khrushchev. By the way, have you heard about the wonderful experiments that Jacob and Monod are doing in Paris?"[2]

"No—it can't go on like this. Would you please go and have a checkup here?"

"Well, maybe. Do you know an intelligent urologist? I have a certain respect for Swedish doctors. But I'll go to one only if he's intelligent."

I telephoned my colleague, Professor X, and hinted that Szilard was a most unusual person who probably wouldn't behave like any ordinary patient. X was very understanding.

We marched in to see X, a highly competent gentleman of the old school with a dignified air. We sat down. Szilard started a conversation apparently meant to test X's intelligence. X looked a little surprised, but played along good-naturedly. After a whole hour, Szilard was satisfied. X was an intelligent person, and Szilard gave his permission to be examined.

I waited outside. When the examination was finished, I was called back in and we had a three-way conference. Szilard had bladder cancer. It was a very large cancer, filling the entire bladder, and it should have been operated on long ago. It was still possible to operate, but it was going to be major surgery. X advised Szilard to go home immediately and have the operation performed in New York, where there were so many excellent surgeons. We thanked him and left. Szilard was neither upset nor dejected. Only for a moment, his ordinarily calm tone of voice changed, and he said in Hungarian: "It isn't particularly enjoyable to grow old."

He didn't elaborate further on the subject, and the meaning of his words is open to several alternative interpretations. After a while, he said: "I don't know if I want to have an operation. I don't trust American doctors. I liked this man. He's intelligent!"

We were back in my lab. Szilard took out his notebook and continued the discussion on immunology as if nothing had happened. After a while he stood up and announced that he wanted to leave. Could we give him a lift? No, he wanted to walk. But what was he going to do?

"I haven't decided yet. If I'm going to have an operation, I'll stay here. If you don't hear from me, call the hotel tomorrow morning. If I've left, that means I've decided not to have an operation."

I called early the next morning. Szilard had left. I went back to the lab, but everything suddenly seemed weary and joyless.

A week later a telegram arrived from New York: "I changed my mind stop arrange operation Szilard."

I obeyed the order and phoned X. He wasn't at all pleased with the message. It was going to be a very major operation. Szilard had no relatives in Sweden. His wife, a professor of public health, lived in the United States. American urologists are the best in the field. Szilard wasn't particularly young. He was a great man whose activities were important for us all. Why, asked Professor X, would he want to come here? I answered X by telling him of Szilard's response to the visit—how he had repeatedly pointed out that he finally had met an intelligent surgeon, one in whose hands he might imagine placing himself. Oh, did he really say that? Well, OK. After a while we agreed, and I sent a telegram with X's response.

One more week passed. I was in Lund for a meeting when Professor X tried to telephone me at my laboratory. Had I received any answer from Szilard about the proposed time for the operation?

We were a young institution, and most of us were still rather inexperienced and unsure of our functions. My secretary, recently hired, was a friendly and helpful girl who understood that her most important task was to make everything easier for everyone. We had no secrets in our little group, and complete openness prevailed. No, she didn't know at all if Dr. Szilard had answered Professor Klein, but there was a telegram from New York addressed to Professor Klein. Should she open it and read it? Professor X was a bit surprised, and said "If you think it's all right, go ahead."

The girl tore open the envelope and read the following message directly into the telephone: "Please explore confidentially the reputation and manual ability of the surgeon as far as this particular operation is concerned Szilard."

When I got back and realized what had happened, I lost all my self-control and screamed at the poor girl until she cried. Then I telephoned X. He was upset, as I expected. No, he didn't want to do it any longer. There was just too much dilly-dallying. He had been against the whole thing from the beginning anyhow. He asked me to tell Szilard that the surgeons at Memorial Hospital in New York were the best in the world.

I was in a really foul mood. It was as if I had committed a great deception. I wrote a long letter to Szilard with explanations and apologies, and conveyed X's advice.

Szilard didn't obey. Instead of going to the surgeons, he went with his wife to a medical library. For several days they read most of the essential information on bladder cancer and its therapy, and then Szilard made his decision. He didn't want to go through with an operation—the risks were too great, and even if all went well he would wind up very disabled. He therefore wanted to plan and supervise his own radiation therapy, but on the basis of new and untested principles. Until then no one had administered a dose of more than 2000 rads of x-irradiation for bladder cancer. Szilard wanted to have 8000, and he wanted to plan the beam angles himself. After running into a great deal of opposition, he succeeded in convincing the doctors at Memorial Hospital to give him 6000 rads.

Szilard marched into his royal suite, his room on the very top floor of Memorial Hospital, completely dominating everything around him. The radiation therapist later said that the name Szilard continued to startle and frighten him even years later. But the therapy worked. The tumor disappeared like melting snow. Szilards's bladder was badly burned by the irradiation, but that didn't matter. A small plastic bag worn on his abdomen discreetly solved all his problems.

The hospital stay became, in a sense, the happiest period in Szilard's life. This lonely man who by his own free will took the most important problems of the world on his shoulders and felt a responsibility to solve them; this man who was more intelligent and thought more quickly than most others; this apparently polite European who didn't take any pains to conceal his impatience when confronted with inertia and stupidity (except in cases where he wanted to achieve a politically important goal); this incredible genius whom Jacques Monod said was as generous with his countless pearls of ideas as a Masai chief who offers his many wives to his friends; this married man who lived without his wife except for a few days each year; this successful but homeless man who always preferred to roam from hotel to hotel in hundreds of cities and to receive his visitors in corners of the inevitable lobbies with his even more inevitable little notebook; this infinitely and unimaginably lonely man suddenly and finally had his own residence, one even more magnificent in a way that the Sun King's Versailles. It turned out that the world was full of friends and admirers who vied to visit him and to send evidence of their devotion now that they no longer feared that stern glance of his, that critical look that demanded their latest results or that probed deeply into the real or imagined significance of their lives and their work. And to top it all, the tumor was completely gone! From his sickbed Szilard could take part in fiery discussions with his childhood friend and colleague Edward Teller. Super-dove Szilard confronted Teller, uncrowned king of all the hawks, on an NBC television program about nuclear weapons. Even Szilard couldn't counter Teller's formidable arguments quickly enough. When he was on the verge of having his back up against the wall, he released his own private little bombshell, modestly and with a trace of a smile: "Well, as you all know, I am now dying of cancer. I have no children, and I don't care a whit what you do with your nuclear weapons. But if I were in your place, I would be very, very worried. . . ."

After that remark, Teller had little chance to get the upper hand.

While Szilard was in the hospital in New York, he succeeded in arranging his long-sought meeting with Nikita Khrushchev, which took place during the Soviet premier's notorious visit to the United Nations. When he inquired of the Soviet UN delegation about such a meeting, Szilard was first told that Khrushchev didn't have time. But suddenly a message arrived saying that, with very short notice, he should present himself to Khrushchev. Szilard's successor at the Council for a Livable World, Matthew Meselson, later described the meeting as a great success. It lasted two hours, and Khrushchev showed great interest in Szilard's ideas on how contacts at the highest level between the Americans and the Russians might prevent nuclear war.

Although the meeting was hastily arranged, Szilard still had some time to think about a small gift for Khrushchev. Having noticed earlier that modern shaving equipment wasn't as readily available in the Soviet Union as it was, in such profusion, in the commercial West, he bought a razor and some disposable blades at the nearest drugstore. He presented the gift to Khrushchev after their talk, together with directions on its use. Khrushchev was as happy as a child—he had never seen anything like this simple and brilliant present.

"I've given you only one month's supply of blades," Szilard said, "but I promise that on the first of every month I'll send another month's supply, until the day comes that nuclear war breaks out between the United States and the Soviet Union."

"Oh, is that so," said Khrushchev. "In that case, I'll stop shaving from that day on."

After his discharge from the hospital, Szilard had several trouble-free years of work. He finished *The Voice of the Dolphins*, consolidated the Pugwash Conferences, and built the group working for "a livable world" (could there possibly be any better way to express a single, unifying goal?). At the same time, he continued his role as a source of ideas, an inspiration, and an interpreter of results for many of us working in biology. When Jonas Salk found himself with a great excess of funds after the problem of infantile paralysis had been solved through vaccination, Szilard advised him to use the money to establish a new kind of biological institution. The Salk Institute came into being, inspired largely by Szilard's plan. It was to be situated on the most beautiful site imaginable for a scientific institution, near the ocean and quite high above sea level, in La Jolla, the "Jewel of California," close to and yet far enough from the diversions of the city. The plan was that, in addition to scientists, a few "resident philosophers" would also be employed: historians, historians of philosophy, poets, mathematicians, and theoretical chemists who would ponder on the origin of life. The design of the building was to promote communication between the basic scientists and their humanist colleagues. It may not have turned out exactly as Szilard had wanted, but there stands the building—full of life, productive, and very beautiful. Szilard died in his sleep from a cerebral hemorrhage after a normal day's work in La Jolla, five years after his cancer was diagnosed in Stockholm. At the autopsy, a burned and scarred bladder without any trace of cancer was found, as expected.

Life magazine carried a big article about Szilard after his death. There was no mention of the Stockholm episode, which was largely unknown to any except those directly involved, but many examples of "Szilardisms." The article also described Szilard's obstinate style in planning and directing his unique and eventually highly successful cancer therapy. I sent a copy to my former secretary, who was by then a happy mother of several children, and wrote that her little mistake was probably a great service to humanity.

In one of the many obituaries, a physicist colleague wrote that Szilard was "an infinitely sweet and somewhat desperate genius." The unique spirit that meant so much to his friends is irreplaceably lost—the void seems just as great twenty years after his death. But his sense of desperation remains and grows greater every day.

5

Are Scientists Creative?

Science! True daughter of Old Time thou art!
Who alterest all things with thy peering eyes.
Why preyest thou thus upon the poet's heart
Vulture, whose wings are dull realities?
How should he love thee? or how deem thee wise?
Who wouldst not leave him in his wandering
To seek for treasure in the jewelled skies,
Albeit he soared with an undaunted wing?
Hast thou not dragged Diana from her car?
And driven the Hamadryad from the wood
To seek a shelter in some happier star?
Hast thou not torn the Naiad from her flood,
The elfin from the green grass, and from me
The summer dream beneath the tamarind tree?

Edgar Allan Poe, "Sonnet to Science"

In the mid-1960s I received a letter from an American colleague, Sol Spiegelmann. He had it in mind to visit us in Stockholm as part of a trip to Europe, and he very much wanted to give a talk about his latest studies.

I was happy about the letter, and gathered my group to inform them. I told them that Sol was one of the most stimulating scientists in current biology, and that his crystal-clear lectures were packed with many unexpected new results from his large lab. We could look forward to meeting this dynamic intellect and scientifically stimulating person, famous for daring experiments that pushed back the frontiers of knowledge and unafraid of unfamiliar territory. His ambition was to solve important central questions rather than to waste time on the minutiae of fields that were already relatively well understood.

Spiegelmann's penchant for formulating daring and at times hair-raising theories was balanced by his talents as a biochemist. His theories irritated some, but many others were stimulated to test them experimentally. Most of the time, when tested, his theories were wrong; but that was relatively unimportant. A somewhat sarcastic French colleague who himself had often proved Sol wrong described him in the following terms: "Sol is often found standing at the top of a mountain. He waves both his arms around, and declaims to all in the valley below, 'Come up here right away—look what I can see from here!' Throngs of people rush up the mountain, but when they reach the top, out of breath, and stand beside Sol, they don't see anything of what he claims to see. But no one denies that it is indeed a very interesting view."

One of my most ambitious students asked why Sol might be visiting us. I didn't know, but I thought that he might have taken an interest in cancer as a biological phenomenon. He would certainly ask a lot of questions about our current work and give us some interesting suggestions. I was looking forward to an exciting lecture and an inspiring evening afterward at my home.

On the way out to the airport, I was thinking about my most recent meeting with Sol, five or six years earlier. We both took part in a Gordon Conference, one of those wonderfully informal meetings of no more than a hundred participants held annually at a small college in New England. There are usually two or three review lectures in the mornings and evenings, but the afternoons are kept free. Informal conferences of this type played an important role in transforming the United States from an inexperienced newcomer into the leader of the natural sciences within a few decades. Discussions on the lawns and in the deserted classrooms, free of ties, jackets, titles, and hierarchy, had a profound effect on scientific communication.

We sat by the lake. Sol was under heavy fire, but he took it in good spirits. He was building dazzling and vertiginous speculations on a dubious experimental foundation. He sat there with his legs crossed, aggressive and happy. His barblike retorts always started with a triumphant "*Look!*" They were followed by a prolonged barrage until the enemy's fort was laid to waste, at least for the time being.

Now the situation was different. Great things had happened in Sol's laboratory since I saw him last. He had developed a new method that was revolutionizing biochemistry and was about to become one of the most important methods in gene technology. Even the molecular biology of today is based more on an extension of Spiegelmann's methods than on any other single technical procedure. He had found that a small fragment of nucleic acid, either DNA or RNA, could "hybridize" (in other words, pair up) with another fragment that contained the same message code. One could therefore use a known gene, for instance a viral gene, to search for similar or corresponding genes in the great cosmos of the living cell. It's like finding a needle in a year's harvest of hay. Today this is the most important everyday method used in all molecular biology laboratories, and is one of every beginner's basic learning exercises. When it was first presented, though, it seemed just as unbelievable as Spiegelmann's earlier work. But this time the discovery was true, easily confirmed, and within the capability of everyone. Even as Sol was flying to Stockholm, his method had already become universally accepted and his great contribution was acknowledged and acclaimed worldwide.

Sol must be very pleased, I thought as I waited for him. I wondered whether his happy irreverence and aggressive, pugnacious spirit had been replaced by the statesmanlike relaxed wisdom of an acknowledged leading scientist. It was going to be interesting to find out.

When he came through customs, I hardly recognized him at first. If it hadn't been for the characteristic short silhouette, I might easily have mistaken this ashen man with dark glasses for someone else. He reacted to my enthusiastic greetings curtly, with a muffled voice and without any sign of delight. During the ride back to Stockholm he sat, quiet and sullen, without removing his dark glasses. I described the schedule for his visit. We would have time for lunch together, and immediately afterward he would give his lecture. Then we would go to my home, where my young colleagues would get together—they were so looking forward to meeting him!

He didn't say a word. It was painful. I described the experiments that we were doing. No reaction. We arrived at the restaurant, and I tried to talk about current events in Sweden. He was totally uninterested. I asked him about some mutual friends. Yes, they were all well—nothing special had happened. And how was his own family? Well. I asked what he'd like to drink. "Aquavit," he answered. The drink vanished in one gulp. "Can I have another one?" he said. He got it, and after a while he wanted still another.

I thought that I should point out to Sol that the alcohol content of aquavit was the same as that of vodka, and that unfortunately he had to begin his lecture in less than an hour. "Never mind," he said. The third drink disappeared as quickly as the other two. He kept his dark glasses on all the while.

Spiegelmann's lecture was as perfect and well presented as ever. I was the only one who seemed to notice a hint of unsteadiness in his movements and a tendency to keep a grip on the table. But in the middle of the lecture, all these symptoms disappeared and his voice regained some of its usual spark. He even took his dark glasses off for a while, revealing a pair of rather bloodshot eyes.

He mentioned his epoch-making new technique only in passing, devoting most of the lecture to his ingenious but farfetched attempts to achieve self-replication of a nucleic acid in the test tube (or, as he put it, to play God). It was obvious that the interests of the lecturer and those of his audience were quite different. Spiegelmann was already tired of talking about his wonderful new technique, the subject of such great interest to most of us. The audience found his experiments, apparently so exciting to him, somewhat trivial and too speculative.

After a few polite but perfunctory questions of the kind that guest speakers in Sweden often have to content themselves with, we went to my home. The dark glasses were back on. All attempts at conversation were fruitless. Oh, well, he's probably tired after his lecture, I thought. My young colleagues will cheer him up! They were already sitting around in a circle in the living room. The air bristled with expectation.

"What would you like to drink?"

"Brandy." That was the first word Spiegelmann had uttered since we left the lecture hall. We tried to start a conversation. No success. We all became rather glum, and our joy vanished like air from a released balloon. Our guest refilled his brandy glass quite frequently, and I felt an intense desire to be somewhere else.

Finally, I decided to try to start a discussion with my young colleagues. I chose the one student in the group who was the least difficult to talk to, and tried to find a subject that would interest him and the others as well: the world of the scientist. How does one get to be a scientist? How should one divide one's time between lab work and reading the literature? How can one avoid depression when work goes badly, and how can one know when one is on the wrong track? When is it time to halt a project? How can one recognize the subtle but crucial difference between constructive patience and obstinate stubbornness? How can one avoid being so seduced by hypothesis that one loses touch with reality?

"The most important question is how to become truly creative," my student said. "Isn't the best recipe to try to understand how creative scientists work, what makes them tick?"

"What nonsense!" shouted Spiegelmann suddenly, slamming his fist on the table.

The whole room went deathly quiet. Everyone stared at the guest. One young girl found the situation so unbearable that she bolted from the room.

Sol took off his glasses and exploded like a time bomb.

"I am dead tired of all the conceited scientists! Who do they think they are? No one would ever have written Bach's music if he hadn't lived. No one could have painted like Picasso. *That's* what I call creativity! But a scientist? If A doesn't discover a phenomenon today, B will do it tomorrow. A scientist is the mirror of nature, and nothing more. His individual contribution, the 'creation' of his experimental results and the rules derived from them—they're all insignificant. Only a charlatan is creative, because he conjures up a world that doesn't really exist."

The monologue gushed out of Sol like a torrential roaring river, occasionally interrupted by small volcanic eruptions.

They always started with the word "Look!" Every "Look!" was followed by a very specific example of unrealistic illusions, conceit, or megalomania, of scientists' total misconception of their own importance, of their inability to view their fleeting, butterfly-like existence as it really is, undefined and interchangeable. *Forte, agitato, fortissimo, decrescendo, quasi dolce con ironia, pianissimo,* and a sudden *sforzato* with another "Look!" that led the way to the next torrent. Now Sol was in top form. His articulation was perfect, without the slightest detectable effect from the brandy. The contrast with his previous behavior was sharp—his images were as colorful as those in his famous scientific talks, but now they had a different purpose. Sol wanted to prove the total futility and vanity of his own existence and that of his listeners. If I hadn't felt so constrained by my role as the host, I would have put on a record of a wonderful English actor reading from Ecclesiastes, with its recurrent "It is all vanity and striving for the wind." But there I sat, numb, perplexed, and fascinated.

The monologue remained a monologue. Sporadic and rather feeble attempts at rebuttal went by unnoticed. It was like trying to control the wind or to change the course of a great, wide river.

Finally, the monologue subsided. The dark glasses went back on. The evening was over.

I later heard that Sol put on the same kind of show, with only a few minor variations, at every stop on his trip: Paris, Rome, and Tel Aviv. Our French colleague took a fresh look at the man on the mountaintop and came up with the following explanation: "As long as Spiegelmann was wrong, everyone was against him. He had suffered a great deal from the failures in his lab and from the hostility of his colleagues. When success suddenly came to him and his results were not only correct but also of immense importance to all of biomedical science, he wanted to prove that it was devoid of importance by going through such a bizarre psychological summersault. If his success was meaningless, then so were his failures and his sufferings."

I don't believe this explanation. It isn't consistent with Sol's past and with his subsequent growth. Long before he became interested in biology, he wanted to become a musician or a mathematician. He often talked about his longing for the imaginative world of mathematics and the fantasy world of music. His scientific hypotheses had always had a certain aesthetic quality that contributed not only to their being accepted so quickly but also to their being discarded when it became obvious that they were based on incomplete and at times erroneous experiments. When he finally succeeded in developing that ingeniously simple method that opened the way for others to attack such important but previously inaccessible problems, it must have seemed to him a kind of anticlimax. It was his practical laboratory technique, and not his ingenious intellectual and artistic musings, that represented the beginning of a new epoch in biomedicine. The virtuoso biochemist, with his impressive intellectual gifts, must have experienced all the accolades with a certain disappointment.

This is probably not the entire explanation. It might be added that Spiegelmann's work after his visit to Stockholm was perfectly in line with his earlier style of behavior. He devoted his last decade entirely to the problem of cancer. All signs of depression vanished. Again he was the man on the mountaintop, waving his arms with the same kind of enthusiasm. He succeeded in inspiring even larger numbers of scientists, who in turn went on to discover many interesting things. The fact that what they discovered was not necessarily what Sol claimed he had found didn't bother him in the least. I saw him again many times, but never again wearing dark glasses. The introductory "Look!" regained its exuberant, challenging ring, and was followed by a torrent of exciting but often incorrect or badly interpreted results from Sol's lab. These results might have had great potential for cancer research, and possibly even for the diagnosis and treatment of cancer, exactly as he maintained, if only they had been true.

Those of us who witnessed Sol's little performance that evening in Stockholm have had a question in the back of our minds since then. Could it be that he was really right? The incident came to mind suddenly and quite unexpectedly as I sat in the Paris studio of the Hungarian-French artist Simon Hantai. I hadn't seen Hantai for 35 years. We had been in our twenties when we had last met. At that time, Hantai was living

with his wife in a studio loft. They had hardly enough money
even to buy bread—all their money went for paint. The studio
was cluttered with his enormous paintings. One whole wall was
entirely taken up by a bizarre abstract creation. "Black macaroni mixed with white macaroni," I thought, without the slightest ability to feel or understand anything whatsoever about the
work. Today same kind of abstractions by Hantai cover entire
walls in the Museum of Modern Art in New York and the Centre
Pompidou in Paris. I am very pleased at his success, but I find
that just as incomprehensible as his art.

I admitted my doubts to Hantai. I wondered exactly what
criteria are used in judging abstract art. How is it that one artist
can be successful while many others who produce similar kinds
of art, at least to the eyes of the average layman, are not
successful? Are artists influenced by current fashions? Are
critics fair, and how do they come to their opinions?

Hantai is a quiet man, and it is rather difficult for him to
express himself. But this time, according to his family, he
became unusually talkative. No, it wasn't at all how I thought it
would be. The appreciation of art today is not radically different from what is used to be. It is just as impossible now to define
simple norms, and we can't now, nor could we ever, predict
which works of art will survive over the long run. Serendipity
and the whims of fashion can play a certain role, but only for a
short while. In the course of time, the only art that can survive
is "true" art. That's how it always has been and how it will be even
for today's abstract art. After a few decades or a century, what
is left is that which is "true" art. Hantai's argument reminded
me of the process of natural selection.

"What do you mean by 'true'?"

He couldn't quite define it, but he thought that the criteria
were probably similar to those used for determining scientific
truth and its survival. I said to him that truth in science is always
based on objective proof and on the repetition of findings by
others. Therefore, scientific truth must be different in some
fundamental ways from artistic "truth," since the latter is based
on an accurate reflection of a subjective experience and/or on
the ability to create a resonant response in the viewer. A painting, whether it be concrete or abstract, that evokes a feeling in

the observer is considered "true" in its own right. Whether the experience of the observer is in harmony with that of the artist is irrelevant.

Despite these differences, Hantai's idea is justified to a certain extent. Are the processes of scientific and artistic creativity really so different in character, since they share as their recurrent theme a common search for truth, even if their methods differ in many ways?

It suddenly struck me that Sol might have been a victim of narcissism, in the Freudian sense—a basic characteristic of all humans, but one whose expression varies in intensity in different individuals. A scientific conclusion becomes true only after it is verified by several other scientists. It is therefore to be expected that the bond between the name of a discoverer and the ultimate surviving truth breaks down quite quickly. It happens today faster than ever before—the process goes on continually right in front of our eyes. It would be odd if it were otherwise. There are more scientists today than during all preceding times put together. Each of us has to make an individual contribution to be able to continue working at all. On the other hand, an enduring painting is forever associated with the name of its creator. Had Sol fallen into a logical trap— had he confused the survival of a name with a fundamental difference in the process of creativity?

Pilate asked Jesus: "What is truth?" I don't know what Bach meant by the short silence that follows those unanswered words in his Passion According to St. Matthew, but it is not inconsistent with our modern-day answer that there are no absolute truths. The quest for truth can be looked upon as a common denominator in science, literature, and art. It is a treacherous road with many pitfalls. Nevertheless, most writers would probably agree with the Swedish author Eyvind Johnson, who wrote: "There are certainly fleeting moments when one can get as close to truth as to fallacy."

There are no milestones posted along the roads traveled by artists or writers. Their progress may appear very much like the metaphor of the Mimarobe in Harry Martinson's epos *Aniara*, in which he compares the lightning speed of a spaceship to the journey of an air bubble in a crystal bowl where a move of a few millimeters takes thousands of years.

Time and space have only relative significance for the artist. His subject is at one and the same time the goal and the means to reach that goal —the omega and the alpha of his creation. The world of imagination and emotion is like a diving bell to the artist, an observation platform which can enlarge with study, experience, and insight but which nevertheless remains a diving bell. At one instant it may seem infinitely wide in comparison with his creative powers, and at the next moment it may seem a hopelessly small prison from which the immortal soul wants to escape. The artist is aware, at some level, that all this is nothing but moods, mirages, and illusions. But a diving bell is what it really is.

The scientist is often unaware that he, too, is sitting in a diving bell, but that is in fact what he is doing. He imagines himself suspended in an infinite sea where he has come to examine some small detail. His ambition is to use his diving lamp to cast light into a dark cave for the very first time. He may suspect that the cave is really deeper than he first anticipated, and that his lamp is weaker than it ought to be, even though it may be latest wonder of technology. If he is fortunate enough to illuminate a small corner and discover something new, he publishes it immediately. Then he goes a little further, until he is pushed aside by others with more powerful lamps, better strategies, or greater perseverance. They confirm his vision or complement and enlarge it. This is the general view that the scientist has of his work. With certain rare exceptions, well exemplified by Spiegelmann, most scientists realize that their contribution is, and should remain, only a small part of a greater, communal effort, one with many participants. Most recognize that their discoveries might, at best, eventually become part of the body of universal knowledge, the shared property of the species, and that they themselves sooner or later will become as anonymous as the monks of the Middle Ages. But this insight rarely makes them depressed. They don't usually hide behind dark glasses or wallow in the depths of the brandy bottle. From the outset, they have accepted this as a self-evident condition of their original contract. The normal human need of most scientists for recognition and reward is usually contained within reasonable boundaries. They might complain about the size of their reward, the poor memories of their colleagues, the ingratitude of their students, or the malevolence of their competitors, but not about the basic principles.

Very few scientists are aware that their so-called objective science can also be thought of as a diving bell. This insight requires some propensity for philosophical thought, some education in the humanities, or an extremely skeptical disposition. But the diving bell of a scientist has room for many more individuals than the solitary cell of the artist or the author. The necessarily collaborative nature of scientific work, the communal nature of scientific ideas, and the inevitable demands of sociological and conceptual adjustments force scientists to function, and want to function, within the boundaries of the spirit of their time. Without any doubt, the attitudes of scientists are formed by a real or imagined urge to reflect reality—to function like a mirror to nature, as Sol the "mirror man"[1] expressed it—but their findings and interpretations can never accurately reflect all of reality. Even the immediate details of a scientific study are, or often become, distorted by being squeezed into a ready-made straitjacket of the reference framework. In the worst case, fortunately quite rarely, the perception of the scientists may become out of touch with the perceptible reality (see chapter 1). Here, and only here, lies the essential difference between art and science. The message of the artist need not be rationalized or documented in some kind of objective protocol, as if it were a cookbook of reproducible experiments. His work need not reflect nature like a photograph. It suffices that the approach of an artist to the expression of truth in language, sound, or form strike a responsive and resonant chord in many listeners or viewers. It doesn't matter to the recipient if it's the *right* resonance for the wrong reason—only that he experience it as "true." For the scientist, however, this is taboo. The charlatan is neither a scientist nor an artist. He is merely a charlatan.

In discussions with some of my friends who have been well educated in the humanities but who are scientifically naive, I have become convinced that the essential difference between art and science is far from clear outside the scientists' own circles. If only subjective truths exist, and if experimental approaches, interpretations, and scientific standards can derive directly from the spirit of the times, where then is the difference between art and science? This is a question that I am often asked. During at least three separate discussions, my friends have mentioned the same name to illustrate scientifically creative reasoning by a humanist, a reasoning that touches the very essence of scientific and also literary creativity. They've all pointed to Arthur Koestler, my illustrious countryman, whom I admire in so many respects. I've responded by saying that Koestler was a fascinating storyteller and, in some contexts with which he was intimately familiar (as in *Darkness at Noon*) an unsurpassed analyst. But his reasoning in matters of biology is completely divorced from reality. No biologist would hesitate to use the word "charlatan" in relation to Koestler's genetic reasoning in *The Case of the Midwife Toad*, and rightly so. Historians have come to similar conclusions about his history-oriented works, such as his book about the Khazars.[2]

Was it the same kind of culturally induced arrogance that prevented Koestler from understanding his limitations when he repeatedly and unhesitatingly overstepped the boundaries between fantasy and reality—boundaries that for outsiders are so subtle and indistinct but for scientists are etched in stone, and which caused Sol Spiegelmann, during the heat of the debate or through his own personal despair, to praise the charlatan as the only creative scientist? Yes and no. Spiegelmann was a great scientist who could also act like a charlatan when he wished. Fortunately, he did so only occasionally, and usually in order to be provocative.

Real charlatans, who often choose the field of cancer as their favorite playground, contend that even Pasteur was misunderstood and at first even renounced. The astute among them can tell many tales, some true and some false, of scientists whose discoveries didn't conform to the spirit of their time and who therefore had to struggle forcefully against the establishment before being accepted. That phenomenon does occur, and often—it is intrinsic to the scientific process. It is difficult to widen the perspective, and even more difficult to break through the "diving bell" of the prevailing conceptual framework. But in our times it is hard to find any examples of scientific geniuses who have to fight against feudal repression. A more appropriate analogy might be the departure of the queen bee from the hive after transferring her role to a new queen. She dies in isolation—she has no individual existence at all. She must be attended by a small but complete social group in order to be able to form a new community. When that task is finished, the new community functions in essentially the same way as the previous one, and within the same constraints.

Scientists can, however, break out from the limits of their period through experiments that can be verified by others. It happens all the time, more often and more quickly today than ever before. They are first confronted by opposition from the establishment, which has a vested interest in the status quo of conventional wisdom and which may feel threatened. It would be odd if it were different. Scientists live under the same psychological rules as all others. But there is one essential difference between science and most other activities. The verification of results by others rapidly and irrevocably breaks down the resistance, regardless of its strength and origin. It collapses like a house of cards, even if that means breaking with the whole established scientific ethos. Shelves of current literature can turn into dusty curios within a few months, leaving behind nothing more than historical interest (see chapter 11).

The word *creativity* can have many meanings. Koestler and Spiegelmann have used it in different ways. Spiegelmann has emphasized a hazardous, relative, and actually quite unimportant criterion: the "survival" of a piece of scientific work. There is no immortality under the skies of eternity. The soul is subjectively immortal, but the dead do not exist. What kind of fantasy world was he talking about, a world of evening newspapers or slick weekly popular magazines? Koestler's "bisociation" comes closer to the essence of creativity. He coined it to describe the ability to move not only within a linear system of values but also between dissimilar sets of standards. However, one might also discuss creativity on a completely different basis.

Early in his career, while still living in the Soviet Union, André Tarkovsky made an unforgettable film, *The Last Judgment*. The film is about an icon painter and his apprentice who, during the medieval period of war, famine, pestilence, violence, and oppression, devote themselves to a single end: to cast a church bell. They are completely absorbed in their task, working as if obsessed. This lifts them from a horrible and objectively hopeless existence, so that their subjectivity is, in Schopenhauer's words, "completely lost in the object" (see chapter 12) and becomes the "pure subject of cognition, without any will" (*reines, willenloses Subjekt der Erkenntnis*). For Schopenhauer, this was the ultimate state that a human could achieve.

Can this fortunate state of existence be attained only in the context of creative work (with those words used in the conventional sense)? Absolutely not! The Hungarian-American psychologist Mihaly Csikszentmihalyi calls this phenomenon "flow."[3] He describes it as a state of complete concentration accompanied by euphoria and an experience of absolute determination. At the same time, the natural split between the person as the actor and as the observer ceases to exist. "Flow" occurs when the task and the ability to achieve it are completely balanced. An overwhelming task creates anxiety. An ability that outweighs the task leads to boredom.

According to Csikszentmihalyi, we all strive to achieve "flow." This force is far more powerful than any rewards or subconscious urges. We want to experience the pleasure of flow over and over again. Children are continuously on the lookout for new flow experiences in their own ingenious but subconscious way; adults generate it more or less consciously. According to Csikszentmihalyi, we all have a need to enrich our lives in this way. Flow can also be viewed as the subjective experience of individual creativity that does not need to have any utilitarian value and is not restricted to art or science.

Scientifically creative work can also induce flow, but a scientist's own definition of creativity bears a strong imprint of scientific or practical utility. By this definition, the subjective value of flow is irrelevant. This utilitarian framework provides a home for many different kinds of scientific activity, and those who practice such science may produce their work in completely different ways. One scientist might look at something that many had seen before but may understand it in a totally new way. A second might develop a new method that demolishes a previously impenetrable barrier to new knowledge. Spiegelmann is an excellent example of this, whether he liked it or not. A third might have a clear intuition about the choice of a soluble problem, and a fourth might uncover a previously unsuspected connection that can then be verified by additional experiments by himself or by others. Even scientists who succeed in inspiring others to join in a collaborative effort on some common problem are often considered "creative" by their colleagues.

Like Proteus, scientific creativity can appear in many different forms. However, in the final analysis, the survival of individual names is totally unimportant. The scientist whose work paves the way toward a major paradigmatic shift is just as important as the one whose name, because of good luck, pure chance, or (rarely) a special talent, is associated with the shift itself. Having great talent is often, but far from always, an asset, but some important and even seminal contributions have been made by scientists whom one could not call talented or even particularly intelligent, try as one might wish. But what then is the most important basic trait of a good scientist? How should Sol and I have answered the student on that unpleasant but memorable evening?

We were too locked into our suddenly inflexible positions to come up with a serious answer. But could one give any answer after having pondered the question seriously? Does scientific creativity have any single common feature that deserves to be identified above all others?

I once heard an anecdote from a Russian colleague whose teacher was Pasteur's last student. When he arrived in Paris, the master was already old and had very fixed and regular habits. A stroll back and forth in the same hallway between 3:00 and 3:30 every afternoon was part of his unvarying routine. One could set a clock by it. His lab staff watched Pasteur from a distance and would never have dared to disturb him, but they could see that he was mumbling something to himself. It seemed as if he was repeating the same words over and over again—something important, something that under no circumstances was to be forgotten. They were all curious. What could it be that Pasteur was repeating to himself all that time? The young Russian was quite small, and so the others asked him to hide behind a curtain for half an hour to listen to Pasteur. It worked. The message consisted of three words: "*Il faut travailler*" ("One must work").

6

Ultima Thule

Fölöttünk fú a förtelmes halál
Above our heads rages the detestable storm of death
from a poem written by the Jewish-Hungarian poet Miklós Radnóti
two days before he died during a Nazi death march in Hungary in 1944,
and found in his pocket after his body was dug up from a mass grave
after the war

I open what seems to be a rather insignificant book entitled *Murderous Science* (*Tödliche Wissenschaft*), written by Benno Müller-Hill, a well-known geneticist and molecular biologist in Cologne.[1] He had taken a leave of absence for one semester to investigate the role German physicians and scientists may have played in the Nazis' "final solution." After searching through documents in archives and libraries, and after conducting personal interviews, Müller-Hill came to a horrifying conclusion: not only did the great majority of German psychiatrists, human geneticists, and anthropologists know about the killings, but the anthropologists actively took part in their preparation and the psychiatrists in their implementation. Furthermore, several groups of scientists used the human material from the death camps for their research.

It all started long before the Nazis came to power. During his imprisonment in Landsberg in 1923, Hitler himself had read one of the most widely used textbooks on race hygiene: *The Principles of Human Heredity and Race Hygiene* (*Erblichkeitslehre und Rassenhygiene*), by Baur, Fischer, and Lenz. The book apparently made a great impression on him, since he subsequently incorporated many of its ideas into his own book, *Mein Kampf.* Ten years later, and seven months before Hitler came to power, Professor Fischer of the Kaiser Wilhelm Institute explained proudly: "Our eugenic movement was in existence long before the Nazi party."

The "theoretical principles" that supported this movement and which were so widely accepted by the scientists of the Third Reich can be traced to three important delusions.

Shortly after the genetic studies of Mendel were rediscovered around the turn of the century, leading German psychiatrists had convinced themselves, without any scientific proof, that psychiatric disease and other kinds of mental disorders are hereditary. In 1920, the psychiatrist A. Hoche and the jurist K. Binding published a book in which they advocated the "destruction of lives unworthy to be lived." At the same time, the anthropologists stated that superior and inferior races could and should be considered different. Jews, Gypsies, Slavs, and Negroes were all regarded as inferior races. The anthropologists and the psychiatrists competed for the asocial individuals whom they considered as belonging to their respective spheres of influence, while the geneticists devoted themselves to a warped distortion of Darwinism and of natural selection. In 1932, F. Lenz wrote that the battle for maintaining the quality of the hereditary endowment was a hundred times more important than disputes over capitalism and socialism. Von Uexküll, a pioneer of ecology, included in his book on human societies a chapter on "parasites in human shape"—the alien races.

When Hitler seized power, the foundation had already been laid for a program that eventually led to the sterilization and later the killing of mental patients. Müller-Hill views the subsequent mass extermination of Jews and Gypsies as a direct and logical consequence of that program. This process began with the dismissal of the Jewish scientists and degenerated to the unleashing of uninhibited "research" for the purpose of racial purification. Eugen Fischer, director of the Kaiser Wilhelm Institute for Anthropology, explained his enthusiasm in the following way: "It is a rare and special good fortune for a basically theoretical science to flourish at a time when the prevailing ideology welcomes it, and its findings can immediately serve the policy of the state." (*Es ist ein besonderes und seltenes Glück für eine an sich theoretische Forschung, wenn sie in eine Zeit*

*fällt, wo die allgemeine Weltanschauung ihr anerkennend ent-
gegenkommt, ja, wo sogar ihre praktische Ergebnisse sofort als Unterlage
staatlicher Massnahmen willkommen sind.*) Later, during the war,
Fischer claimed that Jews and Gypsies represented inferior
biological species. Declaring these groups subhuman made
their mass extermination legitimate. Müller-Hill presents a
gruesome yet dry and matter-of-fact description of how the
scientists in these three specialties—psychiatry, anthropology,
and human genetics—went along, little by little, and how they
finally and almost imperceptibly embraced the "final solution."
A few of them simply shut their eyes to the truth and claimed
that they were doing "pure science." These "internal exiles"
cooperated fully with the much larger group of conscious
activists, many of whom developed into direct executioners.

The murders began in mental hospitals. Since all mentally ill
people were considered to be carrying an incurable hereditary
trait, the victims had no rights. Their fate was decided by the
doctors. Before the war broke out, 400,000 sterilizations had
already been performed on mental patients. When Hitler
began his assault on Poland, doctors were needed for the war
effort, and so the mental patients who were to be sterilized were
killed instead. This was euphemistically called "the euthanasia
program." In a letter dated September 1, 1939, the day that he
invaded Poland, Hitler wrote a letter empowering Reich Leader
P. Bouhler and Dr. med. K. Brandt with the responsibility of
extending the rights of specially designated physicians so that
"mercy killings could be granted to patients judged incurable
after a most critical evaluation of their condition" (*dass . . .
unheilbar Kranken bei kritischster Beurteilung ihres Krankheitszustan-
des der Gnadentod gewährt werden kann*). As the war progressed,
the killing program was expanded to other categories of victims. Müller-Hill notes that the first extermination campaigns
undertaken after the attack on Russia were carried out by young
doctors in white coats working behind the advancing front line.
The "euthanasia experts" of the mental hospitals had the most
extensive early experience in murder at a time when the
extermination program was being expanded quickly to include
Jews, Gypsies, Communists, and other undesirables. The choice
of words is interesting: "mercy killing" instead of execution,

"euthanasia" instead of murder, "final solution" instead of genocide. Mercy killing was said to be "granted." In more recent times, Bruno Bettelheim called attention to the great danger in the use of these kinds of euphemistic distortions. He even objected to use of the word *Holocaust.* One must not convert the straightforward and direct term—mass murder—into an apocalyptic one with mystical overtones.

Müller-Hill searched the many German archives eagerly for evidence of opposition or protest to the "euthanasia program," possibly in order to find solace for his own soul or to be able to maintain his faith in mankind. He found only a few meager examples from priests, writers, and jurists, but not a single written word of disagreement or protest from any psychiatrist, anthropologist, or human geneticist.

We usually think of Joseph Mengele and other criminal doctors as psychopathic charlatans, acting on their own initiative and wishing only to satisfy their sadistic urges. Nothing can be further from the truth. Mengele—or "Herr Kollege Dr. Dr. Mengele," as Müller-Hill prefers to call him in a newly published article[2]—was one of the most respected young scientists at the Kaiser Wilhelm Institute. He had two doctorates, one in medicine and one in anthropology. He began his career as an assistant to O. von Verschuer, who had succeeded E. Fischer after his retirement as director of the institute. Verschuer considered Mengele to be a very talented collaborator who made the most of "the enormous possibilities of Auschwitz" (*die riesigen Möglichkeiten Auschwitz*) for scientific research. According to Verschuer, Mengele had assumed the position of camp doctor at Auschwitz on May 30, 1943, and had provided the Kaiser Wilhelm Institute with "unusually valuable material in the form of well-preserved and packaged human organs." Support for research programs using this material, and also for the scientific instruments at Auschwitz, was provided by funds from the Deutsche Forschungsgemeinschaft (German Association for Scientific Research).

Several other institutes also received "material" for their studies. The director of brain research at the Kaiser Wilhelm Institute, the well-known J. Hallervorden, wrote to one of the extermination camps: "If you are going to kill all these people, could you at least take the brains out and send them to me." He especially wanted brains from people killed by carbon monoxide poisoning, and he regularly sent jars, boxes, fixatives, and packing material for this purpose. The brains were routinely delivered to his institute in a moving van.

Hallervorden's material consisted of 696 brains from people killed in concentration camps or in psychiatric institutions. One was that of a child whose brain was damaged by carbon monoxide poisoning while still in the womb after an attempted suicide by the mother. This child was permitted to grow up and later was sacrificed in the "euthanasia program." Hallervorden thought the case so interesting that he published an article about it after the war, apparently without any moral reservations.[3]

During the war, a great deal of scientific research involving experimental subjects and data from the concentration camps was published, but only in exceptional instances did the journal editors object to any of the studies. For example, the *Zeitschrift für induktive Abtsammungslehre und Vererbungsforschung* (Journal for the Study of Genealogy and Research in Heredity) delayed the publication of a paper, dealing with the inheritance of eye color in Gypsies, in which the author claimed that an entire Gypsy family of eight just "happened" to have died on the same day, thus allowing the eyes of the entire family to be prepared and studied. The paper never appeared in print because of the rapidly approaching end of the war.

Müller-Hill interviewed some of the surviving scientists who had taken part in projects that had been made possible by the "enormous possibilities of Auschwitz." He also talked with several of their former co-workers, assistants, and family members. Not a single one of them believed that his own work or the research of his former directors had any other than purely scientific purposes. Not a single one of those interviewed thought that the people carrying out the studies were anti-Semites. Similar statements had been made by Fischer, Verschuer, Lenz, and others, people who were in fact the driving forces behind the designation of Jews and Gypsies as a different biological species with "foreign proteins." No, it was claimed that they were all serious scientists who personally didn't dislike Jews at all.

The testimony and documents provided to Müller-Hill made it quite evident that the only worries of these "scientists" dealt with the competition among them for recognition or for research funds. In their eagerness to serve the Reich and its ideology, they tried continuously to advance their studies and contribute "improved and more reliable data." The son of the leading race biologist, W. Lenz—himself currently a professor in human genetics—indicated to Müller-Hill that the work of the anthropologists was often based on Fischer's concepts, even though his most important and influential earlier work lacked the slightest trace of scientific proof. Fischer was an anthropologist who believed that he understood genetics even though at the time no anthropologists really understood genetics at all. Lenz Jr. expressed surprise that none of Fischer's colleagues had pointed to the complete absence of genetic documentation in his book. Fischer provided no proof of genetic basis for any of the "racial characteristics" that he considered heritable. According to Fischer's surviving colleagues, they were convinced that his conclusions "had to be correct," and therefore they accepted the concepts of race biology without any critical evaluation—the absence of data and scientific proof was unimportant. In the introduction to the book that served as the first step on the road toward the policy of extermination, Fischer wrote "modestly" that his work was of the *most* infinite significance. Merely infinite wasn't great enough for him, Lenz Jr. pointed out to Müller-Hill.

Lenz Jr. has a picture of his father hanging on the wall next to Thomas Morgan and H. J. Muller—quite a mixed company, Müller-Hill remarks dryly. The younger Lenz feels that the criticism leveled against the Nazi scientists is unfair, since the individuals blamed had been uprooted from their own period of history without any effort being made to understand the thinking of the times. These people were the products of their era and their social class, just as we belong to our own time and to its social milieu. According to Lenz, their attitudes, now regarded as so clearly of malicious intent, were in fact simply based on a "mistaken interpretation of reality, with basically good intentions."

Most of those who used the "human materials" from Auschwitz and other concentration camps remained in their posts after the war, and some were even promoted. Results of their studies were included in their postwar publications and dissertations, but the origin of this material is concealed behind a veil of euphemisms and vague descriptions. Although none of them has ever written about or described his experiences, the War Crimes Tribunals and the Nazi hunters did succeed in flushing some out into the limelight. Most of these men had either been in charge of concentration camps or had operated the extermination program, but their contention that they were merely insignificant cogs in a greater wheel was dismissed. From a reading of Müller-Hill's extensive documentation, one might be inclined to agree with them without absolving them of their responsibility.

The commandant of Auschwitz, Rudolf Höss, wrote his memoirs in prison before he was hanged.[4] He considered himself a very good human being. He described with horror the torture inflicted on the Jews and others before he released them from their torment by planning the very fine and hygienic extermination establishment at Auschwitz! Before his time, diesel exhaust had simply been pumped into vans or primitive gas chambers crammed full with prisoners, causing a slow and agonizing death.

Müller-Hill confirms this account by citing testimony given by an SS officer named Gerstein to the Nuremberg War Crimes Tribunal. The victims could be heard wailing inside the van. Gerstein testified that a physician and professor of public health at the University of Marburg, one Dr. Pfannenstiel, was present during one of the killings. The physician put his ear to the door of the van and remarked: "Sounds just like a synagogue!"

Höss was particularly proud of the poison gas Cyklon B, which he himself introduced—a gas that could kill with 100 percent efficiency in less than 30 seconds. It was possible to keep the shower-like gas chambers quiet and orderly. His descriptions are reminiscent of instruction manuals for the rapid and merciful extermination of vermin or pests. Was Höss more guilty than Fischer, Verschuer, Lenz, and their many colleagues, who lived and died as esteemed scientists? Wasn't it Professor Fischer who declared that Jews and Gypsies were to be regarded as a separate biological species? Hadn't Höss merely taken that "scientifically established fact" to its logical conclusion?

Being the good technician that he was, Höss reacted strongly against Adolf Eichmann's maniacal zeal for extermination. According to Höss, he had frequently pointed out to Eichmann that he couldn't gas more than 3000 Jews per day—that was the establishment's capacity. But Eichmann ignored this protest and sent as many as 9000 people per day. Höss was full of indignation over this reckless overburdening of the facility.

In order to illustrate his own sense of justice, Höss described one of his strolls outside the electrified barbed-wire fence surrounding the camp. Inside, he saw a prisoner carrying a loaf of bread under his arm. Another prisoner crept up from behind, shoved the first prisoner against the fence, killing him, and then took the bread. Höss ordered the immediate execution of the prisoner—he simply couldn't tolerate injustice.

According to Höss, this incident was an example of the animal-like qualities of the Jews. But he silently ignored the issue of why the prisoner didn't have any bread. He found the behavior of the Jews in the Sonderkommandos equally repulsive. These were prisoners who saw to it that other inmates lined up properly in front of the gas chambers and that the entire execution proceeded in a calm and orderly manner. The members of the Sonderkommandos were also periodically executed, as they all very well realized. Höss made no effort to hide his scorn for these people who calmed mothers with lies rather than warn them about where they were really going with their children.

Could one imagine a more humane kind of behavior under these conditions?

More than 40 years have elapsed since the end of the war. Most of the civilized world condemns the Nazi era as a time of complete madness, and one feels righteous and noble in doing so. But isn't the self-congratulatory phrase "I'm OK" just the same kind of wishful thinking that Höss showed within his own set of ethical standards?

The Oxford historian Hugh Trevor-Roper ended his review of Müller-Hill's book with the alarming question of how the mass murder of the Jews might have continued if Hitler had won the war.[5] How much would we know today of the fate of the Jews, the Gypsies, and the other annihilated peoples? How curious would the German people be to learn what had really happened? Would the new generations simply accept the victory and its consequences? One can imagine the answers to Trevor-Roper's questions, but one is reluctant to utter them.

It is just before Christmas 1985, and I'm sitting in a large, old-fashioned apartment in the middle of Cologne. It has a high ceiling, and the walls in all the rooms are lined with overflowing bookcases. Two sweet little children are preparing for Christmas. There are large, rather peculiar modern paintings hanging on the walls, and there is evidence of all sorts of intensive activity in every corner. The rooms show signs of a great deal of reading, writing, and creative children's play, giving an impression simultaneously of organization and disorder. Things considered important are obviously well organized, and unimportant things put aside, much as in my own home.

Benno Müller-Hill is sitting across from me. He tells me about his scientific work. He was one of the first to identify a "repressor protein." This is a signal substance produced by a bacterial gene that can turn off another gene whose function is temporarily unnecessary. The repressor recognizes a distant gene, finds it just as a magnet would find a needle in a haystack, binds to it, and puts it into a kind of long-lasting "sleep." The gene rests and becomes inactive until it is needed, at which time the repressor lets go of its grip. Cells take advantage of this so-called negative gene regulation under many conditions. The role of the repressed gene might be to break down an unusual kind of sugar. This function is not required when that sugar is not available, but it should reappear quickly when the sugar is added. It can then either bind directly to the repressor or inactivate it in another, more indirect way. Ancient viruses have learned that too much virus protein can damage their comfortable existence in the host cell, which is like a "sleeper car" on a train. They use the same repressor mechanism to turn off the expression of their own genes involved in the regulation of the production of the viral proteins, allowing the "train" to continue its journey undisturbed (see chapter 10). But the train is doomed when it meets certain outside problems. If the host cell is starved or exposed to radiation or toxic substances, the virus may shut off its repressor gene remarkably quickly, starting up the viral protein factory and allowing progeny virus particles to pour out like rats jumping off a sinking ship.

Müller-Hill is currently in the process of elucidating one of the most challenging problems in biology. How does the repressor protein recognize the correct gene? How can its amino-acid structure "read" the exact combination of letters in the language of DNA? How can it find its way so unerringly to the exact "address" among all of the hundreds of millions of such letters?

The period when biology could be approached through philosophical speculation is sleeping its eternal sleep now on dusty library shelves. Benno—experimentalist, and molecular biologist—is a distinguished pioneer of the *new* era. He isolates and clones genes in bacteria; he produces and analyzes purified proteins. He has his system well in hand. The repressor protein binds tightly to its unique target gene sequence. These reactions can be carried out on little pieces of filter paper. He is currently changing the letters of the DNA sequence, one by one, slowly and laboriously. How much change can the repressor protein tolerate? When does it become indecisive and wonder whether to stay on or to leave? Which letters are essential? What is it that distinguishes insignificant spelling mistakes from completely new words? How do the two different code languages—that of proteins and that of DNA— communicate with each other?

The telephone rings and Benno's wife answers. After a while she comes back in. A previously unknown Nazi archive has just been found, and Benno is being asked to take a look at it and to find out if there is any new information in it. It contains several thousand pages. Without blinking an eye he agrees.

I am amazed. "How do you do it? How do you find the time? Your scientific work is among the most exciting that I can imagine. You are a professor at one of Germany's most prestigious institutions, you have graduate students, and you know the details of every experiment that they do. It's not difficult to see the sparkle in your eyes when you talk about your research. But those Nazi archives are horrible! Most people would find it unbearable to deal with these monstrosities at all. And you probably have to endure them over and over, without end, and to lurch about in an unending supply of stupidity and egotistical ruthlessness! How can you, of your own free will, jump into this snakepit, this charnel house? How can you voluntarily stick your nose in this rotten dung heap, that proud nose that should instead be sniffing the wonderful fresh air of your science? How can you take so much of your precious time away from that journey into unknown worlds that no one, since the beginning of time, has ever seen? How do you find the strength to deal with that?"

"I have learned to read very quickly," he answers dryly. "Somebody has to do it, I suppose. I recently visited a neuroanatomy institute. According to information in the archives, this institute had the 'finest' formalin-preserved collection of 'interesting brains' from the 'euthanasia' centers. I asked what had happened to the collection filed away under numbers that I found in the archives. Yes, of course it's here! The institute staff thought it was very nice that somebody finally showed some interest in this wonderfully preserved and well-cataloged collection. I asked if many had inquired about it before. No, none at all. As far as they could remember, only one person had ever visited the collection before, and that was sometime after the war."

The brains had been collected by a highly respected professor who is often mentioned in Müller-Hill's book. He had regarded the brains from the euthanasia program as "interesting scientific material" in the same way that his colleague thought that Auschwitz offered "enormous possibilities" for scientific research. Both are now long since gone. Auschwitz is a museum; the atoms of the cremated have permeated the universe and become part of billions of new molecules in living beings of all kinds. Left behind are those splendidly preserved brains, labeled with their precise German labels—a kind of Pompeii, but one that surpasses even the worst possible natural catastrophe in horror.

"Why do you do it?"

Müller-Hill doesn't answer. Instead he gives me one of his earlier books, containing a series of lectures for students: *Die Philosophen und das Lebendige* (*The Philosophers and the Living*).[6] The last lecture is entitled "From the Myths of Race and Heredity to the Auschwitz Culture of Annihilation." It is introduced with a quotation from a ministerial speech given in Bavaria during the Nazi period: "National socialism is politically applied biology." The chapter concludes with these words: "During the last lecture I talked about the past, not quite a generation ago. 'Let it be! That's all history now. Talk about the present, or do some experiments instead,' my well-meaning critics tell me. But I think that's false and misleading. Those who would extinguish or forget the past are in the process of destroying both the present and the future. Only he who understands the past can live happily in the present and help to build a better world."

"I understand in principle what you're saying, but not completely. Exactly what is it that you are warning us of? There is in your book *Murderous Science* a horrible chronicle entitled 'A German Chronicle of the Identification, Separation and Extermination of Those Who Were Different' (*'Eine deutsche Chronik der Erkennung, Absonderung und Vernichtung Andersartiger'*). Your chronicle starts with the rediscovery of the rules of Mendelian inheritance in 1900. It contains the most important facts concerning the publication of books, theories, and official proclamations regarding racial biology, including Fischer's famous 1913 book *The Bastards of Rehoboth*. Then comes the entry of January 30, 1933, the day that Hitler became Chancellor of the Reich. On July 14 of the same year, the law mandating compulsory sterilization for the mentally ill and psychopaths was enacted. On June 25, 1934, Professor Lenz pronounced to

a state committee that only a minority of his German countrymen were 'so endowed' that their unrestricted reproduction
could be considered valuable from the racial point of view. On
October 6 of the same year, courses in race biology for SS
doctors began at the Kaiser Wilhelm Institute for Anthropology
under the direction of Professor Fischer. In 1935, marriages
and extramarital sexual relations between 'Jews and citizens of
German or related blood' became forbidden. In 1937, it was
decreed that all colored German children were to be forcibly
sterilized. On January 30, 1939, Hitler spoke for the first time
about the 'annihilation of the Jewish race in Europe.' The
'euthanasia program' began on September 1, 1939. By October
1939, 283,000 patients had been identified by psychiatrists and
other physicians to fall within the guidelines of the program,
and 75,000 files were marked with a cross, indicating that those
patients were to be executed. In 1940 the use of carbon
monoxide began, and 70,273 mental patients were killed in a
period of nine months. On September 1, extermination by gas
began at Auschwitz and in January 1942, the first gas chamber
was built there. Details of just how the 'final solution of the
Jewish question' was to be organized were discussed at the
Wannsee conference on January 20, 1942. In March of 1942,
Himmler decreed that all selections in the extermination
camps were to be carried out specifically by physicians educated
in anthropology. On May 13, 1943, 'Dr. Med. Dr. Phil.' Joseph
Mengele began his work as camp doctor at Auschwitz. On June
7, 1943, Professor Clauberg, a gynecologist at Königsberg,
wrote to Himmler to say that the method that he had developed
at Auschwitz for the large-scale sterilization of women was
virtually ready. With this new method he could sterilize between 400 and 1000 women per day. At the same time, the
crematorium at Auschwitz reported that its capacity had increased to 4756 people per day. During the summer of 1944,
Doctor Doctor Mengele sent rich stores of research material to
the Kaiser Wilhelm Institute for Anthropology. These shipments included eyes of murdered Gypsies, internal organs of
murdered children, skeletons of murdered Jews, and sera from
twins deliberately infected with typhoid by Mengele himself.

"These and many more such things are described in your chronicle. And then, with a huge sigh of relief, one gets to the entry of May 8, 1945: the war is over! It would have been good to put an end to the chronicle then and there and, if you'll pardon me for saying so, to be able to put your book down. But you don't allow that! The last date strikes like a hammer blow to the head, at least for me: 'On April 25, 1953, Watson and Crick publish their epochal discovery of the three-dimensional structure of DNA, the basis for all of modern molecular biology.'

"What on earth do you mean by that? How could you do something like that to yourself, to me, to all of us? We all know that, in many other contexts, you've said that this is the most important discovery of the century. Your own research and practically all of modern research in biology is based on that discovery. WHAT DO YOU MEAN BY THIS?"

Benno listens to my outburst calmly. As an answer to my question, he hands me an essay: "Genetics after Auschwitz."[7] I look it over quickly while Benno makes us some tea.

A scientist lives in the present, Benno writes. Success and appreciation by colleagues depend on his ability to present new theories and discoveries. A preoccupation with the past can build a barrier between an active scientist and his colleagues. Most scientists can't concern themselves with things that don't lie more or less within their precise areas of expertise. But they quite willingly assume that the human society in which they function is as reasonable as nature itself, or at least nearly as reasonable. They almost always believe that mankind can advance only through an ever-growing exploration of nature.

The German psychiatrists, anthropologists, and human geneticists wanted to play God. They believed that they could decide mankind's future and assume nature's function of selection. From that vantage point, Benno views Auschwitz as a monument to science and technology. What a horrible thought!

Benno agrees that the existence of a state based on law that limits its power to coerce its citizens now prevents the worst from happening. But questions must be raised all the time.

I leave Cologne with my head spinning. I have a deep urge to shake my head, but I don't know in which direction. Yes or No? Both? I can understand some of Benno's conclusions, and in particular I can accept his well- documented evidence that the scientists of the Nazi era could see only a small portion of the reality of their time. Is it possible that their obsession with detail, so understandable and even necessary for scientists, prevented them at least to some extent from understanding what they really were doing?

In his essay, Müller-Hill audaciously extends his argument to the present time. The more scientists understand their science, the less they understand the world around them. This harbors a hidden threat that may lead, in the worst case, to a direct continuation of the "tradition" of Auschwitz and Hiroshima. The natural sciences must not be allowed to lead developments on their own. But what institution of our society should share this responsibility with science? "Everything else that feeds from other sources!" answers Müller-Hill. "Something else— not instead of, but rather in parallel with science. A scientist must be a little 'schizophrenic' and learn to live with two different self-images—the scientific one and the 'other one.' There are no words that can adequately describe the 'other'— they've all been drained of their meaning because they've either already lost their credibility or are on their way to being discredited.

"Those who seek to capture this 'other' by scientific methods will lose it. Those who seek to construct it by scientific methods will destroy it. Those who wish to find it through discourse will make it vanish. Those who try to force it into the open will lose it. It must be lived as an example and cannot be taught. Whoever becomes aware of having it has lost it!"

Again I shake my head, and again I'm not sure in which direction. What in the world is he trying to say? The same as the medieval philosopher Cuzanos? Does he mean that science has no ethical values of its own and must always function within the framework of some other ethical system? But, of course, we have such a system that we can depend on in our democratic world, don't we? Don't our ethics committees ensure that science will not be misused as it has been in the past? Even with my limited experience, couldn't I cite hundreds of examples where medical or biological considerations had to yield to ethical principles? Consider, for instance, the appealing example of organ transplantation. As soon as it had been shown the 25 percent of all siblings are optimal organ donors, it was no longer recommended in any democratic society that siblings serve as donors, since the pressures on the siblings most compatible from the immunological point of view would make truly voluntary organ donation impossible. I could give many more examples. Well, no—it's good that Müller-Hill reminds us of the past, but I simply can't accept that last date in his chronicle, and I can't share his dark view of the future.

The following day I board an SAS flight. I open my Swedish newspaper with the usual calm, secure, and slightly bored confidence in the basic decency and fairness of our society. Just before I put the newspaper aside to submerge myself in the comfort and security of my scientific papers, a small article catches my eye. One of my colleagues, a good scientist and a thoroughly decent person, has recommended that the entire Swedish population be tested for antibodies to the AIDS virus. All those who test positive should be marked with a tattoo or a magnetic signal to allow easy identification.

I have the same feeling in my stomach as I did the night in Jerusalem when I turned the light on in my dark room and saw a big, black scorpion coming toward me.

Quod erat demonstrandum?

Journeys

7

The Campsite

On the road to Kilpisjärvi, 1965. We bought supplies in the last Swedish store in Karesuando, a place where the salesclerks and the regular customers spoke only in Finnish. We put our letters and postcards into the last yellow Swedish mailbox, crossed the Könkämä river by ferry, and started our long northbound drive on the Finnish side. It was late in the evening, but the July sun did not show the slightest inclination to sink below the horizon. During the ferry trip I checked the roof of the car. Everything was in order—the tent, the sleeping bags, the supplies. As head of the family I could relax, since my wife and I had taken care of everything. The three children, twelve, ten, and five years old, all feel safe and very excited. We're on our way to a "foreign country," in our own car! Later in our trip we're planning to go farther north to Treriksröset, the point where Norway, Sweden, and Finland meet. Is it really possible to see how the borders of the three countries merge?

"Can't we go there tonight? Please! . . ."

"No, that's out of the question. We're all tired after our long drive. We'll stop about halfway, pitch our tent, and continue tomorrow morning. The daylight is deceiving—we shouldn't think it's still daytime and just keep going."

After only a few miles, we notice that Finnish Lappland is quite different from Swedish Lappland. Both of them are wilderness. But the Finnish side is inhabited, although the farms and cottages are very far apart. It looks strangely threatening. Who lives here on these desolate farms? Then suddenly I remember:

Large enough the Cape of Suomi,
Wide enough are Savo's borders,
For a man to hide from evil,
And a criminal conceal him.
Hide thee there for five years, six years,
There for nine long years conceal thee,
Till a time of peace has reached thee,
And the years have calmed thine anguish.
Kalevala 35 [1]

How absurd! What strange thoughts in this wonderful, peaceful land!

"Let's find a good campsite. Is it all right over there?" "No, the ground slopes too much."

"Over there, then?" "No, somebody else is there already."

"Let's stop here by the lake!" "No, that's really marshland. We certainly can't pitch a tent there."

"When will we get there?" asks our five-year-old from the back seat.

"Very soon, my child, don't worry."

Everybody trusts the patriarch, the head of the family. His authority is firm, and it is taken for granted that he'll always find the right direction under all conditions. As an example of a father's unerring sense of direction, my wife quotes from Proust—the walk to Combray with the parents. The passage describes long, difficult, and tiring detours:

"Suddenly, my father would bring us to a standstill and ask my mother—"Where are we?" Exhausted by the walk but still proud of her husband, she would lovingly confess that she had not the least idea. He would shrug his shoulders and laugh. And then, as though he had produced it with his latchkey from his waistcoat pocket, he would point out to us, where it stood before our eyes, the back gate of our own garden, which had come, hand-in-hand with the familiar corner of the Rue du Saint-Esprit, to greet us at the end of our wanderings over paths unknown. My mother would murmur admiringly, "you really are wonderful."[2]

As the family authority figure, I can sense a feeling of security in my wife and the two older children in the back seat. They are sure that I know where we are headed. But the five-year-old is more of a skeptic and appears somewhat doubtful. If only she knew how right she is!

Anxiety grows in me like a thriving bacterial culture. The forests seem to be endless and threateningly dark. We haven't seen a single car for at least half an hour.

What a beautiful river! Let's stop here—think how wonderful it'll be to step out of the tent here in the morning!

No, it's too dangerous for children here. Look at those sharp cliffs!

In the stream a cataract fiery,
In the fall a fiery island
On the isle a peak all fiery
On the peak a fiery eagle.
Kalevala 26

My eyelids are getting heavy. No cars in sight, no tents anywhere! Even those desolate farms have now vanished. The car radio doesn't work, because someone "borrowed" the antenna at the last Swedish campsite. Now I really have to stop! But still, only marshland on both sides of us.

Oh, look at that! "What a huge bird!" someone shouts from the back seat. That's certainly true! An unusually impressive bird is flying alongside us like a boat. My wife suddenly remembers that she has put a piece of very smelly cheese up on the roof.

I glance quickly into the rear-view mirror. It's a big, dark bird. Another omen! Ah, yes, didn't it go like this?

Came a bird from Lapland flying,
From the north-east came an eagle
Not the largest of the eagles
Nor was he among the smallest,
With one wing he swept the water,
To the sky was swung the other,
On the sea his tail he rested,
On the cliffs his beak he rattled.
Kalevala 7

But the bird very quickly loses all its mythological connotations. It snaps up a big piece of that nice cheese and disappears. What a relief! Even so, its appearance has added to my growing anxiety. On top of it all, the whole family begins to complain in

unison. "Why didn't you tell us that we still had so far to go? Do
you really know where we're going? You can't keep going all
night, so let's stop now."

The threatened authority figure confidently assures everyone that he really does know what he is doing. A really ideal
campsite is now very close by. Just a few more minutes of
patience! Who has ever seen such a troublesome and impatient
family!

No, we can't go on like this! The road passes over a small
stream, and there seems to be a possible parking spot between
the bridge and the stream. We turn in, park, and get ready to
unload.

Oh no! This is really the worst place imaginable! We are
immediately attacked by a cloud of mosquitoes. The ground is
damp, uneven, and full of stones. The whole area looks like an
inhuman moonscape. The anxiety that I feel over the safety of
my family grows to huge proportions, even as I confidently
proclaim the advantages of this "perfect" campsite. How about
the newspaper reports of those people murdered in desolate
campgrounds? Can't those Russian bears come over here to
northern Finland? Well, at least lynx? How can the five-year-old
sleep, as usual, at our feet on this sloping ground? She'll
probably roll down into the stream during the night. Where is
that mosquito repellant? What should we grab first, the repellant or the tent? "Where did you put things?" "You still haven't
learned how to pack so that we can ever find anything!"

After half an hour, the tent is up. The rest of the family fixes
up the inside of the tent while I check on our hopeless location
again.

I look in through the opening of the tent. Amazing! This is *our*
tent! Everything is just like it was yesterday! Nice and cozy. My
anxiety melts away as if it were snow in sunshine. Everything is
as it should be, as it used to be, and as it is to be for ever. How
could I have been such a complete idiot?

While I slowly slip into the arms of Morpheus, I have a sort of
déjà vu sensation, something in the distant past, something
deeply repressed. Have I experienced this before? No, that's
impossible. We have never been camping in Finland, and
hardly ever in Sweden for that matter. It must be something

else. Could it be simply the difference between the real or imaginary dangers of the external world and the permanence and safety of our home?

Inexplicably, I spend an uneasy and tortured night. Early in the morning I go down to the chilly stream. What was it that I was dreaming about? Was it some close friend or relative that I was trying to talk with, to convey some message, but who turned away from me? Suddenly the memory wells up from the depths of my subconscious. I am nineteen years old. I am in Nazi-occupied Budapest. I have just read the secret report about Auschwitz that my boss at the Jewish Council has shown me. It was written by two Slovakian Jews, the first to have escaped from the death camp and survived. I visit my uncle, a distinguished physician, in one of the better sections of Budapest. He still lives in the same apartment, connected to his medical office. The electrical equipment and all his "ultramodern" medical instruments are all still there—all those things that have been his and his family's pride for so long. The furniture is beginning to look a bit shabby. The joys of my adolescent years, the many books and phonograph records, the beautiful paintings and reproductions, are still in their old familiar places—the acquisitions of a long lifetime and gifts from grateful patients. Every piece carries with it its own special memories. But the nameplate on the office door now has a big yellow star in the upper right corner. That means that the doctor is a Jew and can treat only Jewish patients. Nevertheless, the patients keep coming. They are also wearing yellow stars, but they enter through the same door as always, sitting and waiting just as before and complaining of the same troubles as ever.

"Run!" I say to him. "Drop everything! Run out into the night! It's five minutes before midnight. They're coming to put you in a cattle car with a hundred other people and only one bucket, stuffed together like sardines, to take you to a concentration camp in Poland. They strip you naked and tell you that you're going to have a shower. You are pushed into the 'shower room' with hundreds of other naked people, suspecting nothing. Out of the shower head comes cyanide gas. Your dead body is taken to be incinerated in a big oven. Your gold teeth will be taken out by your former colleagues, the Jewish dentists, before they meet the same fate. The gold in your teeth is melted down to support the German war effort. Your body will be turned into soap, maybe even to be used in the military bordellos where your daughters and your closest friends' daughters are abused until they become sick or pregnant. Then they'll be gassed and incinerated like the rest!"

He stares at me with his mouth hanging open, his neck turning a bright red. "Have you gone completely mad? Do you really believe such horror stories? Have you all been taken in by fright, panic, and hysteria? Have they gotten even to you, a mere messenger in the corridors of the Jewish Council? How can anyone believe that humans are capable of anything like that? Look, here is my office, as it has always been. My patients need me. They trust me. I do a kind of work that is always in demand. My home is full of my own memories and my family's memories. Go tell your wild tales to friends your own age! I don't really have time for this kind of nonsense!"

I go and do as he says. No one believes me. But I know that I'm right.

Did I dream all this? No, all this has really happened. But it was a very long time ago, in another country, under very unusual circumstances. It couldn't happen now, not here! I look around in the tent. My whole family is sleeping peacefully. What is that outside the tent? Possible some friendly reindeer, maybe some Finnish peasants who love to walk alone in the forest, that wonderful and deserted landscape— or maybe something else? Of course! That's what it is. Yes. No? Of course. No. Yes and no. The Gulag is considerably closer to us now than my own home in Stockholm.

8

First Encounter with Africa

In the mid-1960s, our tumor biology lab took the first faltering steps from mouse to man. It began when we found "footprints" of a virus in the membranes of mouse leukemia cells. This raised the question whether similar changes could be found in leukemia cells of human origin. Could that help us to find an unknown leukemia virus?

I was fortunate enough to make contact with an Irish surgeon in Nairobi, Peter Clifford. He was a senior physician in the ear, nose, and throat department at the large Kenyatta National Hospital, where he treated many African children with a malignant tumor called Burkitt's lymphoma. The disease had been described earlier by another Irishman in Africa, Dennis Burkitt, who thought that the unique geographical distribution of this cancer, restricted largely to the tropical rain forest region, suggested that it might be due to a virus disseminated by insects.

For several years, Clifford sent us serum and tumor cells from Burkitt children every Tuesday on an SAS flight. The material always arrived late in the afternoon, and Tuesday nights in our laboratory were transformed into "Burkitt nights." The lab staff took part enthusiastically. Our studies were successful. We published one paper after another together with Clifford, but several years went by before we met. We were nevertheless in continuous contact by mail. Clifford included the children's histories in the dry-ice packages which contained the samples, and in return we fed him the results. It was clear from the beginning that he was an exceptionally effective person. Despite the fact that he was one of the very few ear-nose-throat surgeons in all of East Africa and had an enormous workload, he was passionately interested in this pediatric tumor and had made pioneering contributions toward its chemotherapy.

It wasn't until after several years of collaboration that I had a chance to visit Clifford. I was invited to a symposium in Australia, and decided to make a stop in Nairobi on the way. There was only one flight every other week from East Africa to the west coast of Australia. The schedule was extremely tight, with a stopover of 48 hours in Nairobi. It was to be my first visit to Africa.

I was getting ready to leave from Stockholm early in the afternoon in order to catch the evening flight from London to Nairobi, and the morning therefore became very hectic. In the midst of all my preparations, I received a telephone call from an old scouting friend from my Budapest days. We hadn't seen or heard from each other for more than twenty years, but he sounded as if we had parted only the day before. Now he was living in Buenos Aires and was visiting Stockholm for the first time. He absolutely wanted to meet with me. I explained my situation. Could he possibly ride with me to the airport at Arlanda? There wouldn't be any other chance to get together. He agreed.

During the forty-minute ride between Solna and Arlanda, I was taken back defenselessly to a double world of lawlessness, murder, persecution, and disappearances without a trace: Budapest in 1944 and Buenos Aires in the 1960s. My friend had become something of a legend during the Budapest era. He disguised himself as a Hungarian army officer and drove around in a stolen military car. Together with another former scout from our group, who acted as his orderly, he saved doomed Jews—literally straight out of the deportation train— with the help of false orders for their arrest and execution. Sometimes it was necessary for him to abuse the prisoners in front of the Hungarian guards before they believed him. We regarded him as a mini-Wallenberg. When we met in Stockholm, he was a well-established businessman in Argentina. He described to me the frequent kidnappings direct from the streets as if they were absolutely everyday occurrences. Sometimes they happened close by, even while one was waiting at a traffic light. "But it doesn't disturb our daily life as much as one might imagine," he noted. "As long as one doesn't get involved in politics, one can live very well."

I was astonished. Does one lose the ability to identify with the oppressed when one's own life becomes secure? Can even the most daring partisan convert into an indifferent bourgeois with the passing of years? How may we then sit in judgment of our former Hungarian countrymen who watched the death marches of the Jews with such indifference?

When we arrived at Arlanda, I was really pressed for time, and I wanted to get through the usual airport formalities. But my friend insisted on giving me a small gift, a little flashlight for my keychain— it can be very useful when getting home late at night, he said. Just what I needed!

I cursed quietly to myself as I tried to dig out my keychain, that symbol of the security of my home. After a lot of fumbling around, and barely able to control my irritation, I finally got onto the plane and could at last concentrate on preparations for the following day.

I hardly slept at all that night. Early the next morning I landed in Nairobi. Clifford met me at the airport. My schedule was organized to the very last minute. First we were to go to the hospital for rounds. Then I absolutely had to visit an animal park. We were to drive to Serengeti during the afternoon because we had to be there before darkness set in. We would spend the night there, see the animals, and then return to the airport early in the morning, to make sure that I got there on time to catch my flight to Australia.

The children with Burkitt's lymphoma made an indelible impression on me. I knew them before only as exotic names or meaningless numbers, gradually made more concrete in the form of test tubes of serum, antibody titrations, or specifically labeled cell lines. But now I saw in front of me a beautiful black child with his jawbone swollen to enormous proportions. Despite that, he seemed neither worried nor saddened, but rather indifferent. Or apathetic? But other children who had been admitted with much less severe disease and who still could get around without any difficulty also seemed to be more or less listless. A reaction to the hospital environment—natives suddenly uprooted from their earthy existence and thrown into another world?

Clifford told me that African children never played in the hospital. The hospital staff had given them dolls and other toys and had also tried to involve them in group activities, but without success.

In one bed lay a fourteen-year-old boy who was obviously very sick. His Burkitt's lymphoma was at an advanced stage. Chemotherapy hadn't helped him, and he didn't have many more days to live. His eyes reflected a primal kind of animal suffering, but no anguish.

We continued on to a small children's hospital where, for the first time, I saw the marvel of children cured of their Burkitt's lymphoma. There, in front of me, were children— standing, sitting, running— healthy and spry. I compared them against the frightening and occasionally horrible photographs taken of them on their admission to the hospital—pictures that showed tumors half as large as a head, often with bleeding ulcerations, mutilated eyes and teeth, and distorted lips. But now the swelling was gone. Only small scars remained. A glass eye or false teeth bore witness to their earlier history.

I could hardly believe my eyes. How was this possible? The cell poisons used for treatment of their disease could not conceivably have destroyed all the tumor cells. They don't work with sufficient selectivity, since they also kill dividing cells of normal tissues. This makes them particularly dangerous for cells that must divide continuously, particularly the bone marrow and the mucosal lining of the intestines. Treatment must be limited so that these tissues are not damaged too severely. This means that the doses must be kept at levels below what is required to destroy all tumor cells. The combined therapy with several cell poisons that Clifford used for the Burkitt's lymphoma patients would have been ineffective for the usual forms of leukemias and lymphomas found in the Western world, diseases that are cousins to Burkitt's lymphoma. Despite that, the Burkitt's tumors melted away like snow in the sunshine, never to recur, in a quarter of the children.

Before my visit to Nairobi, I didn't really understand the significance of this word *never*. With his careful and scientifically rigorous use of terms, Clifford had always talked about "prolonged regression" and not about cure. And yet, it looked likely that definitive results had been achieved in these cases. Tumors that recurred after treatment almost always did so within a year. But the majority of these successfully treated Burkitt children had already been at this special private children's hospital for four to five years without a recurrence of their tumor. This indicated that the very last tumor cell might have vanished from their bodies. How could that be, since the treatment itself couldn't possibly have destroyed more than a fraction of the tumor cells? I could think of only one possible explanation: the patient's ability to react immunologically against the tumor cells must have been the decisive factor. If so, the chemotherapy was merely an aid to self-cure. However, this could have happened only if the patient's immune system recognized the tumor cells as "foreign," in about the same way as it could reject tissues transplanted from an unrelated donor. Such a mechanism could, in principle, destroy the tumor cells remaining behind after the chemotherapy had managed to kill only some of them. According to this hypothetical course of events, the relationship between the growth potential of the tumor cells and the reaction of the organism is like a race. In the event that the patient's reaction is the weaker of the two, the tumor "wins" the macabre race and destroys its host and also itself.

From studies of experimental immunology it is known that the rejection process can become inhibited in a number of ways. The patient's own "killer cells" can have difficulty in "seeing" the tumor cells as "foreign," depending on their genetically defective "repertoire of reactivities." But even if the recognition mechanisms function properly as far as the "killer cells" are concerned, the immunological "orchestra" can react in an awkward way—for instance, by mobilizing not only the killer cells but also their opposing number, the suppressor cells. The normal function of the latter is to prevent the immune response from overshooting its mark. In such a pathological situation, it may be difficult for the system to "know" where the mark really is, especially since the tumor cells are continually trying to recover and catch up. Even treatment with cell poisons can affect the balance among the various components of the immune system in a way that becomes detrimental to the patient.

When I visited Nairobi, all these ideas were hardly more than speculation. But they were nevertheless important, since they strengthened our eagerness to pursue our hunt for a virus. Our previous mouse experiments had shown that virus-induced tumors offered the best target for immune rejection of the type suggested by the cured Burkitt's lymphoma cases. The ensuing years have shown that this reasoning was essentially correct.

I asked Clifford how he could keep the children in the hospital for several years. Didn't their families want them back? The answer was both Yes and No. When the children first got sick, it usually took a long time before the families could collect the large sum of money needed for the bus ticket to the hospital. Nairobi is situated high above sea level, and the tropical rain-forest regions where Burkitt's lymphoma is common are far from the city. That is why a jaw tumor often had grown to the size of half a head before the child first showed up. When at last they reached the hospital, the mother often wanted to stay until the child either was restored to health or died. Only after a lot of persuasion did she usually relent and reluctantly return home—in most cases, never to be heard from again. Clifford had to organize special expeditions to search for the families, often after several years. Once, when they finally managed to find a family and announced that there was a healthy child to be picked up in Nairobi, the family at first didn't remember their child. When their memory gradually returned, through the assistance of a grandmother or someone else, they wanted their child back immediately, the same day!

I left the children's hospital with my head spinning and returned with Clifford to the large Kenyatta National Hospital where he headed the ENT department. He saw his remaining tumor patients, most of them suffering from cancer of the nasopharynx. We saw some dreadful tumors of a size never seen in the Western countries. It was later shown that this tumor has an unexpected relationship with the Burkitt's lymphoma virus.

I wondered if Clifford specialized in the care of tumor patients. No, not at all. The majority of the patients admitted to his department of head and neck surgery had various injuries, many of them caused by violence. Crushed skulls were the predominant form of trauma, about ten cases each day. Walking alone in Nairobi alone after dark was definitely not recommended!

Occasionally, Clifford had to admit patients with the most peculiar of all human conditions: the psychosomatic process of dying in a person condemned to death by a witch doctor. This activity of witch doctors was forbidden, but law and reality did not always correspond. A twenty-year-old boy had come in shortly before my visit. He seemed to be very tired after a long trek and said that he felt sick. A physical examination didn't reveal anything abnormal. He was admitted late in the morning. When Clifford returned for afternoon rounds, he could tell even at a distance that something unusual had happened. The hospital is built with fenestrated walls and open terraces. It is usually a lively place that reverberates with all imaginable sounds of happiness, suffering, and various human activities. Now it was completely quiet. Only one voice was to be heard against the resonant background of the dense silence. The patients lay motionless in their beds, many with the sheets pulled up over their faces. The boy was delirious. He prayed, begged, cried, asked questions, answered, threatened, reassured, frightened, comforted, promised. After a while, the experienced ear could distinguish three different voices taking part in the "conversation." The boy's voice begged the spirits of his father and grandfather to intervene for him. He knew that he was about to die, and was afraid of what awaited him "on the other side." The "father" and the "grandfather" tried to console him. They promised to do all they could to assure him of mercy and a worthy place after death. The "trialogue" continued uninterrupted. The boy's expectation of death was unwavering—there was no question whatsoever of survival. After three days of delirium he died, totally exhausted. Nothing abnormal was found at the autopsy.

Modern medicine has no power over the curse of a witch doctor. In the relatively rare instances in which a victim seeks the help of a doctor at all, he would rather not even admit that a witch doctor has put a curse of death on him. He believes that the devil is taking hold inside of him and will tighten his grip even harder if his presence is revealed. Only if the patient doubts the power of the witch doctor does the physician have a chance. If he can increase the patient's uncertainty, then there is a chance to save the victim. Otherwise, not even a respirator will help. It can keep the patient alive for a while, but he will die as soon as he comes off the respirator.

The witch doctor charges for his death curse. His favors are bought for the purpose of punishing real or suspected crimes, such as the theft of a cow. Only the witch doctor who has pronounced the malediction can lift it and thereby save the victim. But it costs much more to annul an established curse than to pronounce it in the first place, and very few can afford it. Nor will most witch doctors with good reputations willingly lift a curse, lest they risk falling into disrepute and suffer the resulting loss of "practice."

Could it be that the Aztecs under Cortez and the Jews under Hitler fell victim to a similar psychological phenomenon? Did they allow themselves to be slaughtered like cattle after resigning themselves to the inevitable, or—an even more terrible thought—after accepting the "fairness" of the death sentence passed on them? Hadn't they been able to differentiate between their own ideal image of a harsh but basically beneficent authority and a cleverly constructed but criminal deception?

My first day in Nairobi was coming to an end. I was stimulated, excited, and overwhelmed, in spite of my sleepless night before. But there was no time to think about that. We had to hurry to reach the animal park before dark. Clifford took me to his English club, the best in Nairobi. It was cool there in that quiet and authentic period piece. All outward signs of the English colonial era remained. There were many black servants in attendance, the gardens were full of palm trees and flowers, and tea was served at precisely 4 o'clock. In the bar sat a few "settlers" from old times, having a few drinks, reminiscing about the bygone days of glory, and prophesying the darkest possible future for the country.

I was given a room at the club where I could shower and change. I went there, locked the door, and got into the shower. When I went back into the room after a few minutes, I immediately saw that someone had rummaged through my clothes, apparently in a great hurry. Someone must have been there while I was in the shower. I checked my wallet. All my dollars and pound notes were gone! The thief had left a small bundle of East African bank notes, all my traveler's checks, my passport, and my airline tickets. He probably specialized in the club and assumed that a foreign guest of one of the members wasn't likely to trouble his host when there was so little proof or time to pursue unnecessary complaints. Absolutely right! I didn't say a single word about the incident, but this, together with the thought of all the crushed heads in the hospital, made me feel rather uneasy.

We had two flat tires on those gravel roads. Clifford was prepared for that, and changed tires very quickly both times. He kept his good spirits while telling me that it was important to change tires very fast, since it was dangerous to be stuck in a car on a desolate road. One could get into trouble even on the major highways. In fact, a car belonging to a government official had been stopped several days earlier on the main road between the airport and the capital. The minister and his chauffeur had been stripped down to their underwear and left on the roadside. It was quite a long time before a car stopped to give them a ride home.

In the evening, we sat on the terrace of the main building of the animal park, several hundred yards from the watering hole where all the animals came to drink. What a fantastic sight! Perfect order ruled there. All the animals knew when it was their turn to drink and when it was their time to leave. The elephants had priority. There was no rush. Even we, as dinner guests on the terrace so far removed and so secure, felt as if we were part of a perfect harmony.

I was given my own bungalow for the night, with strict instructions to stay indoors—I wasn't even to open the door. Absolute house arrest until daybreak! It didn't matter at all, because I was dead tired. I opened all the windows; the mosquito screens seemed to provide security and protection. Then I fell into a deep sleep.

Several hours had probably passed when I was suddenly awakened by a strange sound. Something was stirring outside the screen. It sounded as if something were searching for an opening along the curved wall. I tried to turn on the light. It didn't work! The electricity was out of order!

I panicked. After the theft at the Nairobi Club and all the hair-raising stories that I'd heard, I had only one explanation: a thief must have cut the electrical wires and now had me tightly in his grip. I didn't dare to breathe.

I heard that sound again.

"Who is there?" I shouted as loudly as I could. Silence. I probably had frightened him away.

No, now it started up again. What could I do? I felt trapped.

Suddenly, I remembered my friend from Argentina and his unwelcome gift at Arlanda. I fumbled around for my keychain. I found it and crept carefully to the screen and waited for the sound. As soon as I heard it, I pushed the button. The flashlight worked. Outside stood a colossal bull elephant. I had forgotten that elephants can move about just as silently as cats! He must have been curious about my scent.

The following morning I was on my way again, high above Kilimanjaro. Africa seemed just as strange and dark as before, maybe even more so than ever. But I knew that the Burkitt's lymphoma children and I were irrevocably bound together.

9

The Journey that Wasn't

During the latter part of the 1960s I was invited by UNESCO to give some lectures in a three-week cell biology course in Havana. The purpose of the meeting was to teach graduate students from Third World countries. It sounded very tempting. I had never been to Cuba. But it was impossible for me to leave Stockholm for such a long time, and I wasn't really convinced that I wanted to go anyway. Nevertheless, I answered with some hesitation that I would like very much to participate, but unfortunately I couldn't be away from my lab for more than one week at the very most. To my surprise, my condition was accepted.

There were only two airline connections to Cuba from Western Europe each week. One was a Cuban airline that used turbojet-type Britannia aircraft; the other was the Spanish airline Iberia, which had jets.

The Britannia departed from Madrid at sunset, completely full. We were scheduled to fly over the Atlantic overnight and refuel somewhere in the West Indies in the morning. We were handed newspapers and other literature. The assortment of reading material reminded me of what is usually available to read on planes in the Soviet Union and other Eastern bloc countries. The Cuban newspapers contained long and rather harsh articles about President Richard Nixon. In the cartoon caricatures, the x in his name was replaced by a swastika.

Despite my inadequate knowledge of Spanish, I easily recognized the use of common Soviet terms, most of which often imply something different from their dictionary meanings. Or could the real meaning be hidden among the words? I thought

of the time-honored Hungarian tradition of combining officially sanctioned phrases with "flower talk," a language based
on subtle insinuations between the lines, imponderable double
meanings, and artful linguistic acrobatics that had been developed by some journalists and writers into a kind of intellectual
fireworks display that was understood and enjoyed by the
general public. This use of language has been one of the most
important driving forces for the development of literature, the
theater, and the fine arts during the all-too-frequent power
shifts that have characterized Hungary's history. It is not coincidental that, during the revolts of both 1848 and 1956, the
meanings of the words of poets, writers, and journalists were
first camouflaged and only later became more obvious. But that
could hardly be happening in Cuba, not now, not yet. The
revolution was too young, and it felt most threatened by the
nearby great power, not by one that was geographically so far
away. But then, what did I know about one or the other?

After taking my usual sleeping pill for the overnight flight, I
fell fast asleep, expecting to wake up on the other side of the
Atlantic. But after only a few hours, I was awakened by some
unusually strong turbulence. The captain announced that we
couldn't continue our flight because of a violent storm over the
Atlantic, and that we had to land at Santa Maria in the Azores.

It was about 2:00 or 3:00 in the morning when we arrived at
the international airport of the Portuguese islands I had previously visited as a tourist. It was extremely windy and the rain was
pouring down. The airplane was parked quite far from the
terminal. Nothing was happening. Several hours went by and
the cabin was getting warmer and warmer. The captain announced that he had requested a bus, but for some unknown
reason it was delayed. Dawn came, and the heat in the plane was
almost unbearable. At around 6 o'clock, the bus finally arrived.
At the passport control desk we were told that our plane had
landed illegally on Portuguese territory, that there were no
diplomatic relations between Cuba and Portugal, and that we
were therefore to be detained.

It was only then that I remembered that relations between
Salazar's Portugal and Castro's Cuba were at an all-time low.
Cuba was supporting the liberation movements in the Portu-

guese colonies in Africa. On the other hand, Cuba's relations with Franco's Spain were surprisingly good. Spain had always looked upon Cuba as one of her dearest daughters. Even the enormous ideological chasm between the fascist regime in Madrid and the communist regime in Havana could not seriously affect that bond.

My thoughts went back to the Danube basin, where old national rivalries have always had a far greater influence on the thoughts and feelings of the people than the fleeting political ideologies of the moment. I also remembered the first guest scientist from China who came to visit my laboratory in the 1950s. He was an American citizen and had four American-born children. His father was a government official in Taiwan, and his mother was a powerful party official in Peking. When I asked him about the Mao regime, which was still quite unfamiliar to us, he answered with a trace of a smile: "Every six hundred years we usually get a regime that harbors the illusion of being able to change China."

The police announced that we were to be held in a hotel near the airport, where we were to await the departure of our flight. I waved my Swedish passport and asked if I could go for a swim. Not allowed! "You are a passenger on a Cuban airplane that has landed here illegally. We don't consider you to be in the country and therefore you have to be held with all the others."

Very reluctantly, I had to agree. It turned out that we were in the same hotel where I had previously stayed as a tourist. Then it was summertime, and the hotel was filled with joyful young people. Now it was fall, and the hotel was completely empty. We were the only "guests." I shared a room with a Cuban fellow passenger. We all gathered in the hotel lobby and started to become acquainted with one another and with the crew members, who had changed into civilian clothing. Most of the passengers were Cubans on their way home from South America. Because of the American boycott, they couldn't fly directly home without a detour through Madrid. I met four physiologists on their way home from a meeting in Latin America, a couple of physicists, and several journalists. They impressed me as intelligent, lively, and friendly.

One of the physiologists suggested that we should all sit in a circle in the lobby and tell one another where we were from, what we did for a living, and what we thought of this unexpected situation. He began. No sooner had he said his first few words when the hotel loudspeakers blared out some very loud dance music. The captain asked the porter to turn it off. No, that couldn't be done. "It's forbidden for you to hold any meetings without official permission. We've already called the police."

The blaring music continued so that we could hardly hear our own voices. It wasn't shut off until the police marched in, a whole platoon of them.

"No, you're not allowed to sit around and talk in a group. No more than three of you can sit around a table and talk to one another! You're welcome to dance or socialize in the usual way, but no group conversations! And why are the crew members dressed in civilian clothing? Captain, aren't you ashamed of yourself, socializing with the passengers as if you were one of them? That's not befitting an officer!"

The police officer reminded me of his colleagues in fascist Hungary during the regime of Horthy. The same silly superiority, the same misplaced respect for a uniform, even when worn by a potential enemy. Sometimes even an armband could be enough, as long as it had the right colors. The authority of the uniform could readily transform the most fearful, servile underling or menial clerk into a member of the power establishment in an amazingly short time. In just an instant, obsequiousness turns into arrogance, lowliness into browbeating, humility into vainglory. Could these all be the same basic traits simply reversed? What is the common denominator that fosters a similar kind of behavior in societies geographically and historically so very different from one another ? I wondered whether it could have originated in a similar organization—particularly the oldest and most deeply entrenched authoritarian version, the Catholic Church.

The police under Salazar seemed to have the same notion of respectability as their colleagues in Hungary under Horthy. People had complete freedom to utter all the nonsense that they wanted in Hungary as well. But if they gave the impression that they might entertain some intelligent thoughts, they could easily be suspected of being potentially dangerous.

Well, anyhow, one could put up with that for a while—the storm was beginning to let up. In a few hours, we would be allowed to board the plane again. We were all in a positive mood and felt like old and good friends. We promised one another that we would never return to the Azores.

The airplane rolled out onto the runway and the engines were tested. But—were my eyes deceiving me? One of the four propellers seemed to be rotating more slowly than the other three. I turned out to be right. That propeller stopped a few minutes later. Once again, the plane began to move slowly. Until then, not many others had noticed what I'd seen. It was obvious that we were headed back to the terminal.

The captain apologized and explained that one of the engines had failed to start up properly. We would have to wait in the transit lounge, but he thought that it wouldn't take too long to correct the problem.

There we were, sitting in the same gloomy terminal again. Suddenly, all of us seemed tired and disgruntled. Two of the Cubans took out their guitars and began singing to lighten the mood. They sang beautifully, with an odd mixture of sorrow and confidence, darkness and light, warmth and defiance. They displayed a kind of subtle passion that surprisingly lent some new significance to our existence, which just then seemed so senseless. Suddenly, the song was interrupted by the loudspeaker: "Departure of the Cubana flight for Havana!"

We all shouted for joy and went out to the plane. The members of the crew stood there in their shirtsleeves. The troublesome engine had been completely dismantled! They stared at us and asked why we all had returned.

"Haven't you heard? They just announced over the loudspeaker that we could continue our flight!"

"What kind of idiocy is that? The engine has broken down, we can't fly!"

The captain climbed into the cockpit and called the control tower. "Why did you send the passengers out here?"

The tower answered that it was done on direct orders of the airport manager. He had been very upset. He felt that the two passengers who had been singing had behaved very inappropriately. That's no way to behave in a public place in Portugal!

"The singers will have to apologize personally to the airport manager, or you can all go to hell with your three engines!"

We were all dismayed. Our two performers turned bright red. Never would they apologize to that idiot! The captain talked to them calmly and gently. "We all know that he's an idiot. We really did like your song very much. Please go and do it for the sake of the other passengers."

The singers looked at each other for a moment and said nothing. They went to the manager and apologized formally, after which we were permitted to go back to that obnoxious hotel.

Days passed—one, then two, then three. At first, they hoped to repair the engine, but the local officials would not help at all. The Cubans said bitterly that it would be a completely different story if we could pay with dollars. There were rumors that we were about to get a new engine from England. But bureaucratic problems stood in the way even of that. On the third day, we learned that we were expecting a replacement airplane to arrive from Havana to pick us up as soon as the Portuguese officials would give permission. Nobody had any idea when that would be.

My freedom of movement, which I had previously taken for granted, had suddenly become restricted to only a few square meters. I went over to the terminal five or six times every day to check for any new developments. I watched transatlantic flights land and take off, the same kinds of aircraft that I used to travel in. The passengers came into the transit lounge, wandered around looking at trivial souvenirs and postcards with the same tired apathy I had shown during boring stopovers on long journeys. For the first time in my adult life I suddenly felt like an outsider, and I began to realize what it was like to "peek in from the other side." I tried to stay in touch with my "real world" by writing a review on tumor immunology, but I felt increasingly insecure. What was my real world, anyhow?

At the same time and on a different level I felt like an outside observer, surprised by my rapidly weakening sense of identity. I was acutely aware that my predicament was trivial. I still had my family, my work, my friends, just as before, my country was still there, and my involuntary isolation would surely end after a few days, possibly even a few hours. But I wasn't certain when that might happen, and I was out of touch with my normal contacts. The relatively meager work that I happened to have along was the only link that I had with "my" world. The process of rational thought began to yield to a sense of growing anxiety.

Was it possible that the Cuban physiologists had foreseen all this and might even have been able to forestall it had not the police interfered with our conversations? I remembered some truly life-threatening situations during the Second World War, when simple contact with a friendly person or even the memory of a line of poetry or a piece of music could be crucial for the sense of identity and the will to live. The contrary was also true. One could be dispirited by unfriendliness or even by indifference in casual surroundings out of all proportion to the objective importance of the true situation.

During one of my hopeless strolls between the airport and the hotel, I suddenly remembered what a former schoolmate had told me twenty years earlier. He was majoring in German literature, with special interest in Thomas Mann. During the German occupation of Hungary, he was taken to a military forced-labor camp for all Jewish men of draft age, a place where starvation, typhus, typhoid fever, sadistic Hungarian commandants, and German executioners competed in a macabre contest.

During the summer of 1944 my friend's unit was ordered to perform some emergency repairs on the embankment of a railroad line between Budapest and Vienna in a place that had been damaged by bombs. One day the express train pulled over onto a siding close to where my friend was working in order to yield the right of way to a military transport train. Because the delay was quite long, the passengers were permitted to get off the train and move about. A group of well-dressed gentlemen stepped down from a first-class car, and one could see by their armbands that they were representatives of the Swiss Red Cross. They smoked and talked quietly among themselves, as if it were the most peaceful of times.

My friend stared at them as if they were higher creatures from another world. With a few furtive glances, he quickly realized that the Hungarian officers were taking a rest. He stepped up to the Swiss gentlemen and quickly uttered a few sentences in his best literary High German. He expressed his happiness to see some Europeans again and to realize that Western civilization still existed. Despite all that had happened, he still felt a strong attachment to German-speaking cultural circles. If, by some unexpected chance, he should manage to survive this insane war, he hoped to renew his contacts with any surviving remnants of that culture.

The Swiss gentlemen looked with discomfort and embarrassment at the dirty, unshaven slave laborer who was trying to communicate with them as if from another world. One of them suggested that it was time to board the train again because the weather was turning a bit chilly. They all agreed and left without a word.

I was aware that my involuntary stay in the Azores, when compared with my friend's plight at that embankment, was as a haircut is to a beheading. But I also finally began to understand the comments of one of my medical professors, who thought that every young doctor ought to become seriously ill at least long enough to experience and appreciate the world of a hospital from the patient's point of view.

My own situation couldn't really be compared to that, either. Nevertheless, it seemed as if I had approached the entrance to a long tunnel that led from the depression induced by isolation, through the loss of identity, toward the gradual disintegration of all psychological defenses. I came so close that I could look down the tunnel and at least imagine how dark and deep it really was. Suddenly, I understood the refrain from Phil Ochs' song, made so popular by Joan Baez: "There, but for fortune, go you or I."

In the German concentration camps, the breakdown of personality could reach a special and readily obvious final stage. The people to whom this happened were called *Musulmänner* (Moslems). They were about to die. The unknown prisoner who first coined that expression must have detected a similarity between the emaciation of the prisoners and the world-weary introversion of some devout Moslems. The difference was that the "Musulmann" of Auschwitz hadn't found any new and contented inner life through religious belief. He had become weak and apathetic, moving around wearily and gradually losing more and more of his capacity for rational thought. He had given up all hope and no longer existed as a social human being. All prisoners in the concentration camp realized that there was no return for those who had reached that state, just as there was none for African natives condemned to death by a witch doctor.

Somewhere or other I've seen a study designed to determine if some groups survived the concentration camps better than others. Three such groups could in fact be identified: Orthodox Jews, committed Communists, and Jehovah's Witnesses. Members of these groups could more easily reconcile the unspeakable atrocities of the camps with their overall view of the world. Orthodox Jews are prepared for persecution. They have a profound religious belief and a rigid daily routine that includes rules and prayers for all occasions. They're accustomed to following these rules and to praying openly, but also to praying secretly under the most adverse conditions. Devoted Communists are prepared to spend much of their lives in prison or in labor camps. For Jehovah's Witnesses, world history consists of a series of apocalypses that must be endured before the kingdom of God can appear on earth. The vast majority of ordinary people don't have these defenses. For them, a loss of identity can be life-threatening.

Are we all aware, consciously or subconsciously, that we could easily end up outside the human community, and that we are always running the risk of slipping helplessly into a tunnel from which there is no return?

After four days in the Azores, it was obvious that I couldn't possibly manage to get to Havana in time for the course. I went down to the airport for what seemed the thirtieth time and asked to buy a ticket to Lisbon. No, that couldn't be done. Why not?

"You're not here legally, and therefore you can't leave by ordinary means. You'll have to continue on to Havana."

"Now listen, I have work to do in Stockholm and I have to get back there." The agent turned his back without a word and left.

I phoned my wife in Stockholm, told her about the situation, and then went back to the hotel and fell asleep. In a recurring nightmare that I'd been having for the past several nights, I was back in Hungary with no way out. I knew that we were all doomed, and the only question was: Which day were we going to be put onto the train for deportation?

A hard knock on the door awakened me at dawn. Standing there with a broad smile all over his face was the very same agent who had turned his back on me the day before. Of course I could buy a ticket to Lisbon! In fact, he had brought it with him. The flight was to leave in a few hours. Was there anything that he could do to help make my departure easier? And, by the way, he had an aunt who was suffering from breast cancer. Was there any new kind of treatment?

It turned out that Eva had phoned the officer of the day at the Ministry of Foreign Affairs. He had contacted the Swedish embassy in Lisbon, and from there the message had been relayed to the Portuguese Ministry of Communication, which in turn had contacted the traffic-control director in the Azores. All this had happened during the night. After a few hours, I was "released."

As I sat on that comfortable TAP flight heading for Lisbon, I felt like a contemptible traitor. I had deserted my newly acquired friends, left the plane to its fate, backed out of my commitment to the course, and used influence to buy smiles and kindness. I could not see any extenuating circumstances, except possibly one: to have "experienced the hospital from the patient's point of view" after almost forgetting that the "hospital" exists.

Viruses and Cancer

10

Coexistence between Virus and Man

The relationship between man and his old familiar and experienced
viruses is a beautiful example of peaceful coexistence. The threat
comes from the young, the crazy.

Stockholm, 1984. A huge tome, spit out by computer, is placed
on my desk. It is the longest continuous narrative in the
language of DNA yet seen by the eyes of man. The Cambridge
laboratory of Fred Sanger, a double Nobel Prize winner, had
ground out the 177,000 code letters comprising the genetic
information of a human virus called EBV (Epstein-Barr virus).
Sanger had received his first Nobel Prize in 1958 for the
development of a method to read "amino acid sequences," the
language of proteins. His second Nobel Prize had come twelve
years later for developing a comparable but technically more
difficult method to read the language of DNA. If a protein is
compared with a building, the amino acids can be viewed as the
bricks (twenty different kinds) and the DNA as the blueprint.

The "Rosetta Stone" of biology, which first made it possible
to translate the genetic code into the hieroglyph language of
the amino acids, had already been uncovered a decade earlier,
but it was Sanger's work that enabled us to "read" any protein
molecule and, later, any piece of DNA. The difference between
the two steps is as great as that between being in the British
Museum looking at the Rosetta Stone (the bilingual Greek-
Egyptian record that opened the way for deciphering the
hieroglyphs) and reading the morning newspaper.

Reading and decoding ("sequencing," as it is usually called
in scientific terminology) is performed on relatively small
pieces of DNA. They must be completely purified and ampli-

fied into many millions of identical copies. This process is comparable to copying sentences from a book into an enormous number of replicas. Before the avalanche of information on hybrid DNA technology started, this was possible only with virus DNA, since a small virus contains only a few genes which are obligingly duplicated into as many copies as one might wish. Biotechnology has made it possible to achieve a corresponding mass replication ("cloning," in scientific terms) of any piece of DNA. One often uses a virus, or small and particularly useful portions of a virus, as a vehicle to carry the piece of DNA that one wishes to clone. The small viruses found in bacteria, the so-called bacteriophages, are easy to handle and are therefore often used as basic tools for that purpose. Cloning is carried out by joining, or "ligating," small pieces of DNA together with the portion of the virus structure required for its replication. Such a "construct" then is grown up in bacteria, the usual host cells for the virus. Special, precisely targeted biochemical "scissors," the so-called restriction enzymes that break DNA at very specific places, are then used to cut out, or "excise," the inserted fragment. The code can then be read by Sanger's sequencing method.

Sanger performed his first full sequence on the relatively small bacteriophage called φX174. Other scientists took on other viruses, including some of the small tumor viruses. One was polyoma, a small and highly infectious mouse virus (see chapter 11); its complete sequencing before the end of the 1970s led to important insights into the strategy used by the virus for its action. One scientist working on polyoma gave Sanger the apparently absurd suggestion that he should concentrate his efforts on EBV, the first candidate human tumor virus. EBV belongs to the herpes group of viruses and has a genetic content 33 times that of polyoma. The job took a half-dozen scientists two years, but it was now complete.

I leafed through the impressive bundle of paper and its 177,000 genetic code letters. So this is it. Here are the instructions that can force a cell to subordinate its own growth and replication mechanisms to the interests of the virus. The question of exactly what is in the interests of whom depends, of course, on one's point of view. EBV has coexisted with us and with our ancestors since long before there were human beings. All apes of the Old World harbor this virus or very closely related viruses. New World apes don't have anything at all similar. The EBV of chimpanzees is most closely related to the human virus—more than 90 percent of the code letters are identical. EBV is found in all human populations. It is just as widely disseminated in the highly industrialized nations as in Third World countries. It has managed to infect people in the most remote Eskimo outposts in northern Greenland, in isolated Indian tribes in the jungles of the Amazon, and among the Stone Age people of New Guinea.

Man is no more than 3 million years old. The New World and the Old World apes diverged more than 50 million years ago. EBV must therefore have originated from its nearest herpesvirus ancestors somewhere between 3 million and 50 million years before our time.

EBV is a very successful virus, since it has managed to infect most human beings. Man is also a successful biological species, at least at the present time. The entire wisdom of the virus is inscribed in the book lying there on my desk. Could I read it? Yes and no. It depends on the level of my ambition. Twenty years have elapsed—it seems much longer—since a new Nobel Prize winner arrived in Stockholm, opened a telegram, and saw an unintelligible series of letters. On closer inspection, he realized that the telegram was written in the language of the DNA code. With some difficulty he finally succeeded in deciphering the message: "If you can't read this, give back your Nobel Prize!" Today, every student of biology can crack the code.

Yes, I could read it, but could I understand it? That depends on what the question means. Understand the whole text? Definitely not. Some parts? Well, yes, but at a level similar to that at which a child who had recently learned to speak could understand a reading from Joyce's *Ulysses*. According to some linguists, *Ulysses* is the most "intensive" book ever written in the English language, the one with the richest use of synonyms. Still, Joyce's description of a single day in the life of Leopold Bloom is little more than a shadow of the complex reality. But the virus' DNA is no shadow—it *is* reality, or rather the instructions for the realization of a complicated program for a life process. It has been shaped during millions of years of evolution, through a continuous process of experimentation in which an almost endless waste of time and material was combined with the meticulous precision of stringent selection, as through the eyes of trillions of needles. It is an incessant three-way drama in which the actors are the virus, the host cell, and the host organism. The battle was probably much more severe at the beginning , but it ended well, with mutual adaptation. The present coexistence between virus and man is almost harmonious, a word that can have a rather peculiar ring if you are used to associating viruses with disease. But the symbiotic relationship between EBV and our species is almost as perfect as the coexistence of green algae with fungi in lichens.

EBV has adapted itself to a lifelong persistence that depends on the ability of the virus to remain latent in the lymphocytes without interfering with the normal life of the cell. EBV can also infect epithelial cells located in the lining of the mucous membranes of the throat. We know relatively little about the special relationship of the virus to the epithelial cells, but more about its relation with B lymphocytes. EBV has chosen as its "antenna," or receptor, a membrane component that exists only on the B lymphocyte. A molecule of the virus envelope fits to the receptor just as a key fits into a lock. But the EBV receptor is not merely a randomly chosen membrane component. Its specific function is to trigger the cell to divide and to produce antibodies after it has received certain normal chemical signals. The virus parasitizes this mechanism. Attachment of the virus particle to the receptor releases the trigger. A false signal is sent that favors the life strategy of the virus. As the cell takes its first steps toward antibody production and cell division, the virus penetrates into the cell. There it "sheds its clothes": the nucleic acid loses the protein covering that has protected it against the outside world. The text of the thick book there in front of me, the DNA code message, can now be read by the cell's own "word processor." It functions with great precision, but it is unable to distinguish between the DNA of the cell's own chromosomes and the DNA of the invading virus. But then something unusual happens, something quite different from the strategy of most other viruses. If the cell had been infected with a cousin virus to EBV—the extremely common human herpesvirus (herpes simplex) that causes blisters and sores on the lips and the sex organs—a large portion of the virus DNA would be read. This occurs as a tightly regulated sequence of events, leading to a series of "protein cascades" in which each cascade functions as a trigger for the next. But even the earliest onset of this cascade is tantamount to a death sentence for the cell, which proceeds to rush ahead blindly, making hundreds of new copies of virus protein and thereby digging its own grave. The products assemble to make new virus particles, ready to leave the cell and attack new targets while their former "ship," the infected and dying cell, sinks.

EBV is no less clever in its ability to replicate itself than its more aggressive herpesvirus cousin. But it does know how to wait, to sail as a stowaway rather than sink the ship. When it has settled down in a B lymphocyte, it can prevent the "word processor" of the cell from reading more than just a few of the messages contained in the thick viral DNA book. These newly synthesized "immediate" and "early" proteins work in conjunction with normal proteins of the B lymphocyte to prevent the remaining genes of the virus from beginning their lethal process of protein synthesis. There are up to 100 such genes, as quiet as a row of untapped faucets. But the virus can release itself when the cell meets with danger—the faucets are opened, the cascades begin, new virus particles are assembled, and they free themselves from the dying cell.

By the principle of "live and let live," EBV has provided itself with a protective and stable cell environment. Is it possible that a host organism can derive some benefit from such an arrangement? That is not impossible, but it certainly is difficult to understand what such a benefit might possibly be in the modern world. Today's "amiable" virus, however, may have already begun its adaptation to humans many millions of years ago by protecting its host against a more malevolent predecessor. The original prototype of EBV almost certainly was able to infect many different kinds of cells, and could have caused severe disease. Even today's virus has that potential, but it doesn't make use of it. The conclusive presence of virus receptors on B lymphocytes means that the virus does not enter most other cells. When it is introduced into cells experimentally by microinjection, the result can be violent virus replication and cell death rather than a harmonious coexistence of virus and cell. Today's EBV, adapted to B lymphocytes, may have survived through its ability to protect its host from a more virulent form of the virus, one that could infect all cells. After EBV had lost its ability to infect cells universally, it may therefore have functioned as a kind of live vaccine.

Similar developments are not unusual during a virus' evolution. Some decades ago, the Australians decided to exterminate all the rabbits in their country with the help of a virus. The rabbits had caused a great deal of destruction on that relatively isolated continent, where gentle marsupials had remained in their idyllic and primitive state, never having had to compete with the more advanced placental mammals. The sheep that had been imported had become extremely useful, and now the rabbits (also imported) had destroyed much of the vegetation needed by the sheep. To try to get rid of the rabbits, the deadly myxomatosis virus was to be disseminated, and a strain was selected that was able to kill 100 percent of rabbits in the laboratory within 6 hours. That decision was made by veterinary virologists against the advice of the geneticists, who predicted what the outcome was going to be.

The disease spread quickly. Rabbits died like flies. After several months, no rabbits were to be found, and everyone heaved a sigh of relief. But after a couple of years the rabbits began to return, and now there are as many in Australia as there were before. What happened?

The DNA of the virus, like all other DNA, can change through mutation. When a virus population explodes suddenly in a large number of sensitive host animals, there are many opportunities for mutations to occur. Most mutants survive poorly compared with the relatively well-adapted prototype, which has already been shaped and modified by the previous lengthy process of selection. But one or another rare mutant fares better than the original.

This is exactly what may have happened in Australia: A virus that kills its host after 6 hours survives only if it is transferred to a new rabbit by a blood-sucking insect within those 6 hours. If that doesn't happen, the virus dies along with the rabbit. A virus mutant that kills an infected rabbit only after 12 hours has twice as good a chance to survive, and consequently natural selection nurtures the less virulent mutants. Even very rare mutants of this sort are rapidly selected for and become enriched in the virus population. In turn, they give rise to even less virulent mutants. Gradually, they begin to function as a live vaccine— they induce the production of antibodies that protect the host animals against the more harmful forms of the virus. At the same time, a parallel selection process is occurring among the rabbits. Their immune defense mechanism, a combined ability to generate antibodies to neutralize virus and to mobilize "killer cells" to destroy virus-infected cells, is also influenced by genetic variation. Not all rabbits are equally well armed to react to the foreign proteins present in the outer envelope of the virus or on the membranes of virus-infected cells. A rabbit that can raise an antibody level more quickly or that can mobilize more effective killer cells has a better chance to survive a serious virus infection than an animal with a weaker or slower immune response. As a result of increasing resistance of the rabbit strain and decreasing virulence of the virus, the two reach a balance similar to the coexistence between humans and EBV.

Is it possible that our "tamed" EBV can still behave in a less benign way, or has it completely lost its ability to cause disease? It can in fact cause disease, but only under unusual circumstances. Inoculation of this virus into immunologically "naive" New World apes, for example marmosets, brings on a rapidly progressive and fulminating illness caused by the uncontrolled multiplication of B lymphocytes. These New World apes have never encountered EBV or any of its close relatives under natural conditions, and such an experimental infection therefore meets a completely unprepared immune system.

The uncontrolled growth of B cells is also expressed in tissue culture, where the supply of nutrients is the only limiting factor to the growth of cells. EBV-carrying lymphocytes then grow as "immortalized cell lines" while the EBV-negative cells, present in much larger numbers in the blood, die. In contrast to normal, virus-free B lymphocytes, which have a relatively short life span, the EBV-carrying lymphocyte shows no tendency to age. Despite that fact, or possibly because of it, humans have developed a thorough immune protection against these virus-bearing cells. Our ancestors, or their nearest predecessors, must have been exposed to the same kind of selection process as the Australian rabbits. Our immune system can keep the virus-carrying cells under constant surveillance. It cannot eliminate them completely, but it can prevent them from growing. This thorough protection against EBV does not mean that the virus cannot cause disease, but it does mean that illness occurs only as the result of some biological accident. Immune defenses can break down as a result of congenital defects or in connection with other diseases and infections, including AIDS. Kidney- or heart-transplant recipients must be given immune-suppressive therapy, but this can at times also lead to the growth of the ordinarily harmless EBV-carrying B lymphocytes as a malignant tumor. It is usually easier to treat EBV infections in patients with kidney transplants than in children with congenital immunological defects. Termination of the immune-suppressive treatment leads to rejection of both the transplanted kidney and the EBV-carrying B cells. The patient then resumes dialysis treatment and waits for yet another new kidney. The risk is minimal that the EBV-carrying cells will grow out again for a second time.

Man's ancient coexistence with EBV originally developed under conditions of poor hygiene. Today, the peoples in the Third World and in the poorer regions of the industrialized world live closer to the primal hygienic conditions than the better-protected groups. When hygiene is poor, most infants become infected with EBV, an event usually signaled only by the appearance of antibodies. No disease develops. The perfectly balanced interplay of virus, cells, and host organisms has been disturbed to a certain extent through civilization, however. In families with good hygiene, infants are often protected against infection early in childhood. More than half reach their adolescence still free of EBV. Then they begin to socialize with the opposite sex. In the infected person, EBV begins to multiply in the salivary glands and is secreted into the saliva. This is the basis for what Americans call the "kissing disease" or "glandular fever." The medical term is infectious mononucleosis. The connection of the disease with prosperity is indicated by another popular American name: "the college disease."

Mononucleosis is an immunological civil war. The immune system of a teenager is not as well prepared to deal with the virus-infected cells as the immune system of an infant. The defense response does get underway, but only after a delay and in a rather chaotic fashion. Antibodies and killer cells don't appear until after the virus-carrying B cells have had a chance to multiply. Meanwhile, the lymph glands become swollen, a fever develops, and there is damage to internal organs, especially the liver. A normal immune system usually prevails, and the disease disappears spontaneously after several weeks or, at most, several months. However, the virus always persists in a latent form, exactly as in healthy persons who have been infected during childhood. Antibodies also persist and protect against a new infection. One can get mononucleosis only once.

One of EBV's cousins, Marek's Disease virus (MDV), infects chickens. It is not found in normal, healthy flocks. These birds are therefore "immunologically naive," and the species has not been selected for resistance to MDV. Infection with MDV causes a severe lymphatic disease that can spread like wildfire from the first contaminated coop. This is the only truly infectious tumor that we know. MDV used to be the most severe economic problem for the poultry industry, but an effective vaccine is now available.

Not all chickens are equally susceptible to MDV. There are relatively resistant flocks in which the disease runs a more limited course. The resistant chickens mount an antibody and killer-cell response more quickly than the susceptible chickens. In principle, MDV therefore should be capable of reaching just as harmonious a balance with the chicken as EBV does with the human, particularly if a large proportion of the chickens become infected over prolonged periods of time. The prehistory of the virus-host interaction is thus important in determining the ability of a virus to produce a disease. The same virus can be innocuous in its natural host but lethal for a closely related host whose cells permit virus replication and which has never had any previous contact with the virus in nature. We can illustrate this with a third herpesvirus. The natural host species of this virus, called herpesvirus saimiri (HVS), is the South American squirrel monkey. The virus has been isolated from every squirrel monkey examined, whether it lived free in the jungle or captive in a zoo. It doesn't cause any illness in its natural host. It was therefore quite astonishing when it was found that experimental infection of marmosets resulted in 100-percent-fatal malignant growth of virus-infected lymphocytes. HVS doesn't affect B lymphocytes, as does EBV; rather, it affects the other major category of lymphocytes: the T cells. Normal "killer cells," which among other things have the task of protecting organisms against virus infections, are also recruited from the T-cell category. Accordingly, HSV infection assaults the defense system from within. One might say that the police recruits suddenly turn into outlaws.

What are the decisive differences between resistant and susceptible host species? What happens early in virus infection? In a comparison of the antibody production in response to virus infection in the marmoset and the squirrel monkey, we found that both produced about the same amount of antibodies. But whereas the squirrel monkeys achieved a high level after only 4 or 5 days, the marmosets responded much more slowly. When they finally raised their antibody levels, after 10 to 14 days, it was already too late. By this time, the number of virus-infected lymphocytes had already exceeded the number that can be kept under control.

The difference between the speed of the antibody response in the two species is probably not the only determining factor. Natural selection is a powerful mechanism. It acts on the basis of genetic variation, and it favors the survival of increasingly resistant variants irrespective of the mechanism responsible for their resistance. In some cases, the balance between the disease-causing agent and the host is reached only after a long series of "experiments" by nature. A good example can be found in the rich tradition of Australian research in parasitology.

Australian sheep are sometimes infested with a nematode worm. There is usually only one well-encapsulated adult worm living in the liver, and its presence is not usually harmful to the sheep. The worm sheds its eggs continuously into the intestine, thereby disseminating them in nature to infect other animals. But the eggs cannot establish themselves in the original animal, where the mother worm has had time to stimulate the development of antibodies and other defense reactions. The adult worm is not harmed by those defenses, since it has produced a thick coat that keeps the host cells and the antibodies at a safe distance. But the immune reaction is able to protect the sheep from reinfection with naked eggs that have not yet had time to produce a capsule. They can do so only if they develop undisturbed in a newly infected animal, before antibodies can be produced.

If one introduces the parasite eggs experimentally into goats or other species that are closely related to sheep but are not natural hosts for the worm, different situations can arise. In one extreme case, an animal may respond too slowly or may, from an immunological point of view, mount an awkward and poorly coordinated reaction. Such an animal cannot protect itself against the eggs. It soon becomes filled with worms and dies. This interaction has a poor survival value, both for the goat and for the worm. In another extreme case, the unnatural host may mount a highly effective reaction to expel the worm. The host animal's immune system "recognizes" the parasite, possibly because the parasite proteins happen to resemble other proteins against which the species has been selected to respond. If the response to expel the parasite does not completely eliminate it, the way is open for an optimum adaptation, as far as the worm is concerned. This result is called "concomitant immunity." The encapsulated parasite protects the host against its own offspring. In the long run, it serves the interests of both.

Experimental attempts to adapt parasites to new host species have shown that such optimally adapted parasites could appear by mutational events that can start from either one of the two extremes mentioned before: the generalizing or the rejection-prone type. The mutations are quite rare, but their strong selective value gives them an advantage. The mutation frequency can be increased by radiation or by chemical agents that act on the DNA. One can increase the frequency of mutation by modifying the DNA. As we approach the end of the present millennium, we have gained control over most infectious diseases, at least in the industrialized world. Control in Third World countries is hampered more often by social and economic factors than by medical or biological limitations. One of the most severe virus diseases in history, smallpox, has been completely eradicated. The basic discovery that led to this great achievement was made in 1796 by the Englishman Edward Jenner, who observed that milkmaids fared better during smallpox epidemics than other people. It was a harmless virus closely related to smallpox, the cowpox virus, that protected them. Jenner's brilliant insight, based on the similar appearance of smallpox and cowpox blisters, was immediately translated into practical action. He began his cowpox vaccination against smallpox by infecting himself and then his own family. It took almost 200 years before the smallpox virus could be removed from the list of existing species. It remains now only as a kind of museum object in a few high-security laboratories. This triumph is the first of its kind in the field of preventive medicine. Others on the list of the greatest human epidemics, such as plague and cholera, have not yet been entirely eradicated, but they have been driven back and confined to restricted "niches." Tuberculosis, only a few generations ago one of the most humiliating and widespread of all diseases, has abated dramatically because of better living conditions and antibiotic therapy, and it is no longer life-threatening. Vaccination has succeeded in preventing polio in large parts of the world. The parasitic diseases malaria and bilharzia (schistosomiasis) still affect a large portion of the human population, but they could be controlled if the involved Third World countries had the necessary resources, determination, and knowledge. We are thus approaching a situation where infectious and parasitic diseases ought to be eliminated from the list of humanity's greatest problems.

But just as we were about to heave a sigh of relief, there appeared a new, totally unexpected infectious disease with frightful effects: AIDS.

If EBV can be considered a wise old virus, HIV (human immunodeficiency virus), the causative agent of AIDS, is the aggressive youngster who doesn't know what he's doing. Some information indicates that the virus spread recently from Africa, possibly with a stopover in Haiti. It appeared in the Western world first among homosexuals and later among intravenous drug abusers and people who had received transfusions of contaminated whole blood or blood concentrates.

Where had the AIDS virus been residing earlier? Presumably it had existed in some isolated group of humans, or even in apes.

HIV is related to HTLV-I, the first recognized human leukemia virus, found most commonly in southern Japan and the West Indies. Some humans infected with HTLV-I develop an unusual form of acute leukemia called ATL (adult T-cell leukemia). Contrary to the common acute form of childhood leukemia, ATL affects adults and develops from the T lymphocytes described earlier in this essay.

HTLV-I is found commonly in Japanese apes but doesn't seem to cause any illness among them. It is possible that the virus spread from apes to humans in southern Japan, but this cannot explain its presence in the West Indies. According to an alternative hypothesis, the virus migrated in the opposite direction several hundred years ago, transported by seafarers from Portugal to southern Japan. All this is still very unclear, but a great deal is known about the present routes of the spread of the virus. The risk of infection with HTLV-I is relatively small. The virus is transmitted primarily, and possibly exclusively, by living cells, usually from man to woman or from mother to infant.

T cells infected with HTLV-I usually remain quiet and inactive, very much like EBV-infected B lymphocytes. But the human immune system cannot keep them under tight control, as might be expected from the limited experience of humans with this virus. The immunological defense system breaks down sooner or later in about 10 percent of infected people, after which the infected T cells grow out to produce a progressive, fatal leukemia. The reasons for this collapse are not known. The ability of the immune system to recognize the viral proteins is influenced by genetic variation among the infected people. Because of that, some seem to be able to cope better than others. But the ability of the immune defenses can also be influenced by environmental stresses. And, as in the case of the squirrel monkey and its herpesvirus, there is the added special problem of the T cell as the policeman running the risk of turning into an outlaw. However, HTLV-I spreads through the body less quickly that herpes saimiri. Leukemia probably appears only after a large number of T cells have been infected.

The AIDS virus, HIV, affects the same type of T cells as its "cousin," the leukemia virus, but its cell-killing effect is greater than its growth-stimulating effect. It can remain quiet in cells that happen to be in a "resting" phase. But when the host cell is activated to carry out its normal immunological functions, for instance as a result of contact with an antigen, the virus can begin its own protein production. This becomes devastating because T cells play a central role in the immune system, like a spider in the middle of its net. In addition to their specific functions, including the "killer" activity described earlier, T cells also produce a series of different chemical "signal substances" that help to mobilize other cells of the immune system. The awakening of the HIV virus torpedoes all these functions. The T cell is forced to produce more virus in place of its normal products, and the newly released virus infects still other T cells—a vicious circle. It is especially ominous that the normal functions of the immune system help the virus to further decrease the capacity of the immune system, such as the response to other infections, thus increasing the ability of HIV to lessen the immune capacity further and finally to destroy it. This also explains why common and usually trivial infections hasten the immunological collapse. The disease erupts when so much of the T-cell system is destroyed that the patient becomes defenseless, even against infections that are are usually harmless.

As is true of HTLV-I, the AIDS virus is transmitted more easily by living cells than by cell-free material. Homosexuals are particularly vulnerable, partly because of this cell-bound virus transmission and also partly because of the many other infections that burden their immune systems. Similar problems are also found among intravenous drug abusers.

Is this century going to end with a horrible new plague? The increasing spread of the AIDS virus among heterosexuals (so far mainly, but not exclusively, through prostitution), the transmission from infected mothers to their newborns, and the alarming increase of AIDS-infected drug users are all important warning signals. The unique ability of HIV to destroy the cells whose participation is required to maintain normal immunological functions presents a special, but not insurmountable, difficulty. The biggest uncertainty has to do with the continued evolution of the virus. New virus mutations are occurring constantly. What kind of mutant is going to take the lead? Will it be easier for a more virulent mutant to spread to new host organisms, or will the disease-producing ability of the virus be a handicap in competition with more restrained mutants? No one can answer this question. In the long run, it does seem likely that a less virulent, more innocuous variant will thrive, as in the cases described above. But that might take a long time. Measures to lessen the spread of the disease through increasing knowledge and awareness of its modes of transmission seem to be most important. The silver lining in this dark cloud is the relatively limited infectivity of HIV in comparison with many other viruses.

Epilogue

It was discovered in the 1950s that viruses could reproduce in cell culture. Before that time, the world of viruses was known only through its disease-producing representatives. Many new viruses suddenly appeared when apparently normal tissues were put into culture. No one knew that these viruses existed. They had been hiding in mucous membranes, lymph glands, kidneys, or—surprisingly often—in the brain. Most of them probably would never have made themselves known if scientists had not put cells that just happened to harbor them into artificial environments in which they could reproduce, still within the cells but out of reach of the control systems of the whole organism.

When these facts first became known, during the latter part of the 1950s, a symposium was held in New York entitled "Viruses in Search of Disease"—a paraphrase of Pirandello's play *Six Characters in Search of an Author*. The thought that every virus had to have its "own" disease was still firmly entrenched. Since that time, scientists have increasingly come to realize that harmless viruses are the rule and disease-producing viruses the exception. Peaceful coexistence has a great adaptive value for both parties.

Where, then, do viruses come from? Can they have originated from the cell's own genes that have somehow broken loose? According to Salvador Luria, one of the founders of modern virology, a virus is a collection of genes that has developed a machinery to help it move from cell to cell. Another theory postulates that viruses originated from free-living microorganisms that gradually lost their capability for independent replication and began to parasitize the corresponding mechanisms of their host cells (see the following chapter).

The question of the origin of viruses will probably never be fully solved. It is likely that they have evolved along several different pathways. There is one definition that probably comes closer to the truth than most other concepts of viruses: "A virus may be both a regressed parasite and a partial cell genome that has become infectious, depending on which phase of the evolutionary history of its genetic material we are observing. It may have been both things at different times."[1]

11

The Tale of the Great Cuckoo Egg
(From Tumor Viruses to Cancer Genes)

An age or a culture is characterized less by the extent of its knowledge
than by the nature of the questions it puts forward.
François Jacob

Microbiology celebrated its first great triumphs around the
turn of the century. The world of the microorganisms had been
opened. Previously unknown disease-causing agents were identified in quick succession, vaccination and serum treatments
became parts of the therapeutic arsenal, and many contagious
diseases were defeated. Life expectancy increased, infant
mortality decreased, and the future seemed to be getting
brighter and brighter.

The microscope played an important role in the early development of microbiological science. The size of bacteria was
very close to the limits of resolution of this instrument, but they
were nevertheless large enough to be identified and classified.
Filters were produced whose pores were smaller than bacteria,
and material containing microorganisms could be sterilized by
filtration.

But there were exceptions. Certain pathological agents, invis-
ible under the microscope, passed through the filters. They
were called "filterable poisons," or "viruses." (*Virus* is Latin for
"poison.") But were they living things or chemical poisons? Not
until several decades later did it become clear that a virus was
a new type of organism whose size was below the magnification
capability of the ordinary light microscope. The complex and
geometrically beautiful world of viruses became visible when
the electron microscope came into use during the 1940s. It
then became understood that viruses are different from all

other microorganisms in *one* fundamental respect. Bacteria, algae, yeast, and other microscopic microorganisms are independent, single-cell creatures that can replicate in an unlimited fashion if the external world doesn't limit their access to nutrients or restrict their salt concentration, temperature, and other factors. The frequently rigorous demand of microorganisms for a "correct environment" reflects a long evolution and a resulting high level of adaptation to their surroundings. Virus particles, however, cannot replicate all by themselves. In contrast to bacteria, they are not adapted to the outside world, but rather to a parasitic existence within an involuntary, unsuspecting host cell. Instead of the thousands or tens of thousands of genes in bacteria, large viruses have only a hundred or so, and small viruses have only a few—at times only two or three. But they are used with unbelievable economy. In contrast to the "junk DNA" of higher organisms (see chapter 12), the hereditary machinery of viruses contains nothing unnecessary.

All viruses replicate at the expense of cells. The strategies different viruses use are based on very different principles, but they do share some common features. The genetic instructions used by viruses are written with the same nucleic-acid code words as the genes of all other living things on our planet. They use the same basic letters as you and I, the house plant in your room, the fungus growing in your refrigerator, and the goldfish in your aquarium. The person whom you love the most and the agents responsible for the most dreaded diseases speak the same genetic language. But there is *one* vital difference between viruses and cells. The instructions of the cell are stored in large, long DNA molecules. Cells of the more highly evolved species ("eukaryotes") pack their DNA into microscopically visible rods, the chromosomes. They sit in the well-demarcated nucleus of the cell and direct everything from there. But the DNA doesn't all function at the same time. Small fragments, or strips, containing short sets of instructions copied from single genes, are sent out and find their way to the protein-producing factory of the cell, the cytoplasm. They direct the production of components needed by all cells for their energy and metabolic needs or for the next cell division that will generate two daughter cells. Some determine the production of "luxury goods" made only by highly specialized cells, such as pigments, hormones, snake poison, or silk. The small "messenger strips" are also written in the language of nucleic acids; however, they don't assemble into the large, clumsy DNA molecules, but rather into small, mobile molecules of RNA. The languages of DNA and RNA use the same code words, except for some small spelling differences. The transfer of information from DNA to RNA, a process of rewriting in the same basic language, is therefore called "transcription." Interpretation of the RNA language into the concrete reality of protein building blocks is called "translation." The protein product is built as the RNA instructions are translated (for more on this subject, see chapter 12).

Whereas no cell can function without both DNA and RNA, a virus needs only one of them. There are DNA viruses and RNA viruses. Smallpox, shingles, warts, chickenpox, hepatitis, and mononucleosis are caused by DNA viruses. RNA viruses are responsible for influenza, measles, mumps, German measles, polio, and AIDS. The RNA or DNA of the virus is able to take over the instructions for protein synthesis in the infected cell. The cell executes the instructions of the virus to its own detriment—it produces viral proteins instead of its own cell proteins.

DNA and RNA are fragile molecules. Blood and other body fluids contain enzymes that attack all nucleic acids that happen to be released from their ordinary location inside the cell, sometimes by cell death. The enzymes quickly degrade nucleic acids to their building blocks. The code words fall apart to their individual letters and the text becomes meaningless. The same enzymes also protect our body against any invasion by foreign nucleic acids whose instructions might cause genetic confusion or, worse still, assume command. But that is precisely what viruses do. How have they been able to circumvent these safety devices?

Viral nucleic acids are surrounded by a protective coat that prevents contact with the degrading enzymes. The coat also contains a protein that helps the virus particle attach itself to certain susceptible target cells. Salvador Luria defined viruses as "elements of genetic material that can determine, in the cells where they reproduce, the biosynthesis of a specific apparatus for their own transfer into other cells."[1]

Where does the viral nucleic acid come from? There are two different theories. One views viruses as extremely well-adapted microorganisms that have, during their parasitic evolution, lost all of the machinery needed for an independent existence outside cells, much as whales have lost their legs. At the same time, they gradually evolved the ability to utilize the protein-synthesizing apparatus of the cell for their own selfish aims. According to the other theory, viruses arise from within the cell. They may have started as ordinary cell genes but subsequently acquired an ever-increasing degree of autonomy. As Luria sees it, the transition from a somewhat autonomous gene to a virus would have occurred when the viral nucleic acid became capable of producing a protein coat to protect it against breakdown and to facilitate its transfer from cell to cell.

Neither of these two theories claims that the virus "learns" anything. Like all living organisms, viruses have evolved on the basis of random variations in their nucleic acids, a never-ending experiment with new mutations and the selective survival of forms that are most fit to cope under the given conditions.

What happens when a virus particle infects a new cell? The virus coat attaches to a "receptor," a molecule on the outer membrane of a susceptible cell. The true function of this receptor within the normal existence of the cell varies from case to case, but certainly has nothing to do with its coincidental ability to bind the virus particle. It is the virus that has adapted itself to use a certain component of the membrane as its receptor. The choice has not been by chance. The receptor can be especially appropriate as a "gate" or it can activate the cell's metabolism in a way that promotes virus reproduction.

After its attachment to the receptor, the virus has to introduce its nucleic acid into the interior of the cell, and its protein coat now plays a less important role. Some viruses use their coat as a kind of disposable syringe that is discarded after its nucleic-acid contents have been emptied into the cell. Other viruses fuse with the cell membrane. The proteins of the virus coat float around like icebergs in the fatty sea of the membrane, just like the cell's own proteins. Their presence serves to promote the release of mature virus particles from the cell at a later stage of infection. But there is a long way to go before that, and the interim events are directed by the viral nucleic acid. After the virus has penetrated to the interior of the cell, the nucleic acid is stripped of its attached virus protein and can then get down to work. Most of the viruses insert their genetic instructions in the form of short RNA strips into the protein-producing factories of the cells and force them to make virus proteins in a rigorously programmed sequence. At the same time, the synthesis of cell proteins is blocked. The earliest virus proteins form an apparatus to make more copies of the viral nucleic acid—often many thousands of new copies. This is followed by the "late" phase of infection, when the protein components of the virus particles themselves are produced. The finished proteins and the newly made viral nucleic acids assemble into the symmetrical structure of a virus particle. When the new viruses are finally released to try their luck in new cells, the original cell is already dead or dying.

At the beginning of the twentieth century, when the contours of the enormous structure that was to become modern biology first came into view, half a century before viruses were seen in the electron microscope and several decades before it was known what a virus really is, scientists began to wonder if cancer might be caused by a virus—or, in the language of the time, whether it might be transmitted with a cell-free filtrate.

Groping Steps in the Morning Mist

In 1911, a young researcher at the Rockefeller Institute, Peyton Rous, decided to approach the problem experimentally. Through a stroke of good luck, he chose chickens with large tumors of their connective tissues—sarcomas—as the experimental animals. He ground up the tumors, prepared cell-free filtrates from them, and injected them into newly hatched chicks. The results of the experiment were positive! The same kinds of tumors developed at the sites of injection after several weeks.

It was neither an accident nor a sign of rigid conservatism that Rous didn't receive the Nobel Prize for this discovery until 1966, when he was 86 years old. The fact is that it took half a century before the significance of the discovery was understood.

At the very beginning, Rous' cell-free transmission of the chicken sarcoma aroused a great deal of interest. But by the 1920s the phenomenon had become something of an insignificant curiosity, since attempts to repeat the experiment in mammals gave negative results. To be sure, there were recurrent and scattered reports that even mouse and rat tumors could be transmitted with cell-free filtrates, but subsequent experiments showed that these studies involved experimental errors. The "cell-free" filtrates were not really cell-free, and the developing tumors therefore arose as a result of cell transplantation rather than as a result of new viral induction of a cancer.

At the beginning of the 1930s, one of Rous' colleagues at the Rockefeller Institute, Richard Shope, reported that he had succeeded in transmitting a benign wart, or "papilloma," into rabbits with a cell-free filtrate. The resulting warts occasionally changed into cancer, but only relatively rarely. Rous became interested in this phenomenon of conversion of bad to worse, which he called "progression." He pointed out that the development of the cancer was only indirectly related to the virus in this case. In contrast to the truly virus-induced chicken sarcoma, the rabbit virus caused only the preliminary stages of cancer. This was probably the reason why Shope's discovery was also put on the shelf and forgotten.

The next important observation came from mouse geneticists at the Jackson Laboratory in Bar Harbor during the mid-1930s. Through persistent inbreeding and selection, they had produced pure strains of mice that differed from one another in their frequency of cancer. The purpose was to study the inheritance of cancer under well-controlled conditions. They set up brother-sister matings between the offspring of cancer-bearing and cancer-free animals and then selected for and against cancer over many generations. It worked. High-cancer strains of mice showed more than 90 percent incidences of the type of cancer that had been selected, including strains with high incidences of breast cancer, leukemia, lung tumors, and adrenal-gland tumors. The frequency of all cancers among the low-incidence strains was reduced to less than 0.01 percent, whereas in an unselected population about 5 percent of mice get cancer.

These experiments led to the important conclusion that there is no general cancer disposition, but rather that there are different genes that can influence tumor development in various tissues. In order to study these genes more closely, crosses were made between high- and low-incidence strains. The methods of classical Mendelian genetics were then used to examine the frequency of cancer in the first and second generations of hybrids (the F1 and F2 generations) and in the offspring of "backcrosses" between F1 hybrids and the parent strains. The results of these studies could be expected to give clues to the number of genes that can influence the appearance of a certain tumor, and to the dominant or recessive properties of those genes. The studies showed differences with different tumors, but a totally unexpected observation was also made: the hybrid offspring from a cross between a high-breast-cancer strain and a low-cancer strain developed breast cancer at a relatively high frequency if the mother belonged to the high-incidence strain and the father to the low-incidence strain, but the offspring had a low incidence of cancer if the opposite was the case.

How could the mother transfer high susceptibility to cancer? There were three possibilities: through the cytoplasm of the egg, through the blood circulation during fetal life, and through the milk.

Three young scientists at the Jackson Laboratory were assigned the job of studying these three routes. John Bittner had the easiest task—the milk—and got a positive result. Newborns from the low-incidence strain got breast cancer if they suckled from a foster mother of the high-incidence strain. However, the frequency of these cancers did not approach the 90 percent of the foster mother's own strain; it went no higher than about 30 percent. The evidence concerning the influence of the milk was nevertheless conclusive, since the reverse experiment gave the opposite results. When pregnant, full-term females of the high-cancer strain were delivered by Caesarean section and their newborns received milk from a low-cancer-strain foster mother, the cancer frequency of the offspring decreased dramatically—from 90 percent to 20–30 percent .

Bittner was convinced that the milk must contain a virus that increased the risk of breast cancer in the mouse, but that this virus was neither necessary nor sufficient for the development of breast cancer. He preferred to use the term "milk factor" instead of "virus." The modern name for the virus is MMTV (mouse mammary tumor virus). When Bittner was asked a couple of decades later why he didn't call it a virus from the very beginning, he answered that he didn't want to challenge the scientific world, which had earlier concluded that cancer has nothing to do with viruses: "If I had called it a virus, my grant applications would automatically have been put into the category of 'unrespectable proposals.' As long as I used the term 'factor,' it was respectable genetics."

O tempora, o mores! Two or three decades later, one might have run the risk of *not* receiving a grant if one had talked about genetics rather than viruses. But both attitudes are based on unrealistic expectations and are therefore equally indefensible. I'll return to this point later.

It took fifteen more years after Bittner's discovery before the pendulum began to swing. However, the change in the attitude toward viruses didn't depend on any major differences between the phenomena discovered between the 1930s and 1950s; rather, it depended on how the results were presented and interpreted. Bittner was not only cautious about the word *virus.* His interpretations were based on scientific analysis and a certain puritanical modesty. The breast cancers in the mouse were certainly malignant tumors of a mammalian species, but the milk virus could not produce tumors in all infected mice and its removal did not prevent the appearance of all tumors. The virus could therefore not work alone, but only in combination with other factors. The detailed analysis of this system, which took twenty years and occupied many scientists, is still among the most beautiful in the history of cancer research, and the difference in the susceptibility of the sexes to cancer was one of its springboards. Only females got breast cancer, even in the high-cancer strains that carried the milk factor. But after castration and treatment with female sex hormones, breast cancer occurred even among the males, indicating that hormonal factors played an important role. Inherited factors contributed as well. Their significance has been suggested by the experiments described above, in which the introduction of the milk virus to low-cancer strains increased the frequency of cancer, but to a level significantly lower than the frequency found in the high-susceptibility strain. Freeing the high-cancer strain from exposure to the milk virus lowered the incidence of cancer, but it still remained far above the incidence in the low-cancer strains.

What are these genes that affect the development of breast cancer in the mouse? They can be divided into three main categories. Some of them increase or decrease the ability of the milk virus to multiply in the mouse by directing virus replication in the cells or by affecting the ability of the mouse to respond immunologically to the virus. Still another category of genes affects the hormonal environment. Their existence was first suggested through a comparison of the cancer frequency in different virus-bearing high-incidence strains. Castration prevented cancer in one strain but not in another. Another comparison between two genetically different high-cancer strains showed that females which had given birth and had nursed several litters developed breast cancer with the same high frequency in the two strains, whereas females which had never been pregnant developed cancer in one strain but not in the other. The analysis of these differences led to the discovery of new genes that affect the complex interplay between the pituitary gland, the adrenal glands, and the ovaries, an interplay that affects susceptibility to breast cancer.

The genes in a third category were especially interesting. These genes were detected thanks to the development of transplantation biology, a field also based on the use of inbred strains of mice. The most important "law" of transplantation biology says that tissues can be transplanted freely *within* a highly inbred (homozygous) strain but are rejected when transplanted *between* strains. The first-generation (F1) hybrids derived from a cross between two inbred strains are "universal recipients" for tissue transplants from both parent strains. With this information as a starting point, normal breast tissue was removed from F1 females whose mothers were from a high-cancer strain and whose fathers were from a low-cancer strain. Then normal breast tissue was transplanted from the cancer-prone parental strains onto the right side, and from low-cancer strains onto the left side, of the same F1 female. The tissue developed completely normally. The tissues from both parent strains were present in the same female recipient that carried the milk virus and whose hormone production was stimulated by continuous breeding. Cancer developed 10 times more frequently in the tissues of the high-cancer strain than in those of the low-cancer strain! Therefore, genes in the breast tissue itself must have affected the probability of cancer development. Similar cancer genes with specific target tissues were identified later through the same technique, even in cases of leukemias, tumors of the adrenal glands and the lungs, and other kinds of cancer.

How could a single high-cancer strain contain so many genes that contribute to malignant transformation in so many different ways? This is a reflection of the often-underestimated strength of the selection process. The mouse geneticist Hans Grüneberg used to draw a parallel between the high-cancer strain and a race horse. Everyone knows that a race horse can be bred through selection. But how do the genes responsible for the horse's ability to win races function? Do they affect the bones, the muscles, the ligaments, the blood circulation, the heart, the nervous system, or the psychology of the animal? It is easy to realize that all genes that control these systems can play a certain role. No single gene alone is either necessary or sufficient for the horse to win a race, but each gene can increase or decrease its probability. The animal breeder who selects for a certain complex trait promotes all genetic variations that can contribute to such a trait, whether they exist in the original population or whether they arise later through mutation. But every one of these genes affects the relevant trait only in a very indirect way.

The analysis of breast cancer in inbred mice was the first stimulus to cancer biologists to reevaluate their field. The notion that *every* disease should have a *single* cause was useful in the early studies of infectious diseases, but was counterproductive in the field of cancer. It was gradually realized that tumor development is a complicated process, or series of processes, the probability of which was affected by a number of factors. At first, only a relatively small group of cancer biologists appreciated this new insight. However, from an historical perspective, it was actually quite fortunate for the continued development of cancer research that the scientifically correct approach wasn't more widely appreciated in the beginning. Paradoxically enough, the great breakthrough in the cancer-virus field during the early 1950s came through more traditional concepts. It is easy for us today to see the faults of this approach and laugh at the naive expectations and biases involved. But we can scarcely fail to admire the combination of determination and insight that finally produced the results that everyone was hoping for, and to be delighted with the fortunate circumstances that favored their theoretically misguided but experimentally correct development. The conclusions and the interpretations were wrong, since they reflected the urge to satisfy wishful thinking and justify false hopes. The experimental findings were correct. One should not underestimate the great power of self-deception. The enthusiasm resulted in an enormous investment of resources and manpower. Although the scientists didn't get the answers they were looking for, look at what they got instead!

The Pendulum Swings

An American scientist of Polish-Jewish descent, Ludwik Gross, published a sensational paper in 1950. In contrast to countless negative reports in the literature up until that time, Gross reported success in transmitting mouse leukemia with cell-free filtrates. He prepared his filtrates from a leukemia that had appeared in a high-leukemia strain of mice and injected it into newborns of a low-incidence strain. After 3 months, the recipient mice developed the same kind of leukemia as the donors. Gross interpreted these results in a completely different way than Bittner, owing partly to the differences in their backgrounds and their basic ideas. Gross was a relatively recent immigrant and more or less unknown in the United States. He was a low-ranking worker in a hospital laboratory far removed from the important research centers. He was trained as a physician and had only a marginal familiarity with biology. But the most important difference lay in Gross' absolute conviction that cancer was due to a virus, possibly only a single virus. It was this unshakable belief that gave him the courage and strength to persevere in lengthy experiments in a field that many others had already abandoned and where no one expected any positive results.

Gross' success depended on three factors. His selection of one inbred high-leukemia strain provided him with a virus-carrying tumor cell. The process of prolonged selective inbreeding had produced a strain that carried both the virus and the genes that promoted the development of leukemia, in a way reminiscent of the interaction between the milk virus and cancer-promoting genes in the breast-cancer strains. But Gross didn't know this; he thought that his discovery would pertain to all leukemias. His crucial choice of a donor strain was therefore fortuitous to a certain extent. However, the second factor was deliberate. The recipient mice were newborns, less than 24 hours old. He correctly anticipated that the undeveloped immune system of the newborn animal would increase the chances of a successful infection with the virus. The third factor involved a piece of almost unbelievable luck. Among all the strains of mice available to him, Gross happened to choose the only one whose genetic makeup permitted this virus, eventually to be called the Gross virus, to induce leukemia. Had he chosen any of the other strains, his experiments would have been just as negative as all those that preceded them, and the history of cancer research would have taken a different turn.

It took five years before Gross' result was confirmed by a "respectable" cancer scientist. This long time can be explained only by the fact that Gross was not taken seriously. I can remember a comment by a leading cancer biologist during a coffee break at a conference during those difficult years for Gross: "If this man is honest, he has the incredibly bad luck of not giving that impression."

Gross was indeed honest, but because of the general lack of confidence in him, no one bothered to repeat his experiments exactly—to make filtrates from the same kind of leukemia, to use absolutely newborn mice, or to infect the only susceptible recipient strain. It was the developer of the high-leukemia strain himself, Jacob Furth, who bothered to repeat the experiment exactly as Gross had published it. Furth's short paper, in which he reported that he could fully confirm Gross' results, triggered the most radical revolution in the history of cancer research. The attack proceeded on a broad front. Filtrates were made from all possible sorts of tumors and injected into newborn mice all over the world. Lo and behold! In only a year or so, a dozen new tumor viruses were isolated. Positive results were not limited to leukemias, but included many other kinds of cancer in numerous species: hamsters, rats, cats, apes, and previously untested species of birds.

Today we know that the Gross virus belongs to the same family as the Bittner virus, and that its tumor-causing potential is just as limited, depending on the genetic makeup of the recipient animal. Its presence alone is not a prerequisite for the development of mouse leukemia, and its absence is no guarantee against leukemia. But this was not generally understood during the 1950s. It was more common to imagine that viruses caused cancer through a direct effect. In other words, it was thought that viruses transformed normal cells into malignant cells in a single step. Two scientists at the National Institutes of Health, Robert Huebner and George Todaro, formulated these ideas into a specific hypothesis, according to which the hereditary material of the virus contained an "oncogene"—a gene that can cause malignant transformation. The oncogene would be distinct from genes that directed virus multiplication, which they called "virogenes." Oncogenes and virogenes were seen to be transmitted from parent to offspring through the germ cells in the same way as the virus was transmitted "vertically" in the high-leukemia strain that Gross happened to choose for his studies.

One of the originators of the oncogene concept, the virologist Huebner, went even further. He suggested that *all* forms of cancer arise through the activation of latent virus-carried cancer genes. This theory holds that even chemical substances that can cause cancer (such as those found in cigarette smoke) can produce tumors through the activation of latent virus, and that "spontaneous" tumors occur when the concealed viral oncogene becomes activated through mutational events.

The general enthusiasm for the concept of cancer viruses captured the attention of the most important source of funds for cancer research: the U.S. government. In 1971, Congress passed President Nixon's proposed "Cancer Act." It was hoped that the problem of cancer could be overcome in a short period of time by the investment of many billions of dollars. Most of the new funding went into the Virus Cancer Program, whose main goal was to isolate the presumed human tumor viruses.

But nature cannot be conquered by force. The correct but wrongly interpreted experiments led to false premises. Virus-induced tumors do indeed exist, but they constitute a small minority of human tumors—most forms of cancer have nothing at all to do with viruses. That, of course, doesn't mean that the huge investment of funds was in vain. But science progresses along its own, often quite tortuous paths.

The rapid development of molecular biology and especially of DNA technology was one of the most important by-products of the Virus Cancer Program, comparable to the great impact of the Apollo project on computer technology. Much of the basic knowledge about the genetic code and the structure and function of the genetic material came through this essentially well-intentioned effort to solve the cancer puzzle. In its later phases, the revolution in molecular biology has also benefited cancer research. But its most important contribution so far has been the discovery of the true oncogenes—the huge cuckoo egg in Dr. Huebner's viral bird's nest. But we must search the nest before we can find the cuckoo egg.

The development of virus technology during the 1960s made it possible to purify large amounts of the Rous, Gross, and Bittner viruses, and many other viruses, and to analyze them biochemically. The Gross and Bittner viruses belong to the family of "retroviruses," or RNA tumor viruses. They generally contain three genes: one that codes for the "core" protein of the virus particle that protects the viral nucleic acid, one that codes for an enzyme that allows the viral nucleic acid to integrate itself into the genetic material of the cell, and one that produces the protective outer coat of the virus. Only these three genes were present—no fourth "cancer" gene. In contrast, the Rous virus, which produced tumors much more quickly than the Gross and Bittner viruses and could even transform normal cells into cancer cells in tissue culture, contained a fourth gene. Through ingenious genetic experiments, it was proved that the latter was a true cancer gene, since its presence was both necessary and sufficient for the rapid cancer transformation effect. The gene was called src (pronounced "sarc" and representing a contraction of *sarcoma*, meaning a malignant tumor of connective tissue).

Where did this src gene come from, and did it have anything to do with the life cycle of the virus? The astonishing negative answer to the latter question came through genetic experiments. Forms of the Rous virus that contained a src gene inactivated by mutations grew normally but could no longer transform normal cells into cancer cells. The src gene was therefore not necessary for virus replication. But why did the virus contain such an insignificant and possibly burdensome extra genetic "component"? A virus usually "packs" and "trims" its genetic material with great economy. The origin of the src gene was clarified when the techniques of gene cloning were developed so that one could amplify fragments of nucleic acids into a large number of copies. The src gene was cloned and tagged with a radioactive marker. This "hot" gene was then used as probe to search for corresponding DNA code words in normal tissues, in cancer tissues, and in other viruses, with the help of the nucleic hybridization technique (see chapter 5). To their great surprise, researchers found src-like sequences in many normal tissues, and not only in chickens. All vertebrates, including humans, contained related genes, which came to be called c-src to distinguish them from the viral v-src.

How did the src gene end up in the virus? Retroviruses carry their genetic information in RNA; hence the alternative name "RNA tumor viruses." All viruses of this family go in and out of the cell's DNA, but before they can integrate into the cell DNA the viral RNA must first be transcribed into the language of

DNA. That is accomplished by an enzyme produced by the second gene of the virus, as described above. The DNA form of the virus, called the "provirus," inserts at random sites into the cell's DNA sequences. It remains hidden there and is carried along during the cell divisions exactly as if it were a cell gene. When it is time to produce new virus particles again, the provirus DNA is rewritten into RNA. This RNA has two functions: to produce more RNA to pack into virus particles and to function as instructional (messenger) RNA to be translated into virus proteins. This shuttle traffic in and out of the cell carries certain risks. The viral DNA can grab a piece of the neighboring cell DNA and carry it along. It can well afford to be so careless—a small fraction of defective virus particles among many millions doesn't make very much difference. Cell DNA often takes the place of a portion of the virus DNA, and virus particles that carry cell DNA therefore become defective. In the Rous virus, cell DNA has been added to the complete virus DNA, but even in this case one might expect that the unwieldy, unnecessarily burdened virus particles would would lose in competition with the smaller, normal particles. But Dr. Rous got in the way with his syringe. His goal, tumor production for the sake of science, favored the accidental defective form that happened to pick up a cell gene whose function could satisfy his wishes.

But what kind of a gene can sit quietly in all cells yet become capable of transforming normal cells into cancer cells when it is carried along by a virus?

Rous virus isn't unique. Approximately forty other RNA tumor viruses with a similar direct transforming ability have been isolated. They are usually called "acute" RNA tumor viruses, in contrast to the Gross and Bittner viruses and others, which are called "chronic." Acute RNA tumor viruses have been isolated from chickens, ducks, mice, rats, cats, dogs, and apes. Every virus has its own favorite cell which it prefers to transform. Susceptible cells occur in all three major kinds of tissue: connective tissue, skin and mucous membranes, and blood-forming organs. Virus-induced cell transformation can therefore result in corresponding tumors, sarcomas, carcinomas, and leukemias.

As in the case of Rous virus, all acute RNA tumor viruses contain one extra gene derived from the genetic material of the cell itself. They all carry their own characteristic oncogene, although in some cases the same oncogene has been picked up two or three times in different virus isolates. In all, around twenty different virus-borne oncogenes have been identified. They all have their own names, which, like src, generally consist of three letters. The collective name is v-onc for virus-carried oncogenes and c-onc for their cellular counterparts. The best-known individual oncogenes encountered in today's cancer-biology literature are called myc, myb, erb, ras, fes, mos, and sis. What is known about these genes?

All known oncogenes, like src, have been conserved by evolution—in other words, they can be found in all vertebrates, with only minor variations. Some can even be traced to invertebrates, such as the fruit fly and baker's yeast. Highly conserved genes usually fulfill some "housekeeping" function that every cell must carry out in order to live, such as respiration, energy conversion, or cell division. What kind of housekeeping functions can the oncogenes have in the normal cell?

The answer to this question is far from clear, but certain general patterns can be discerned. The oncogenes can be divided into four functionally different groups that nevertheless have one common denominator: they all regulate cell division in one way or another.

The process of cell division is triggered by chemical signals called "growth factors," which are very similar to hormones. They are secreted by one cell type and can influence cell division in another type. The target cell captures its special signal molecule with the help of a similarly specialized "antenna." This receptor, located on the cell membrane, releases a signal, which is then transferred through an unknown number of relay stations inside the cell and finally reaches the cell nucleus itself, where special DNA-binding proteins stand ready to set the finely regulated DNA replication process into motion.

The four groups of oncogenes correspond to the four steps of this process: the growth factor, the receptor, the signal-relaying functions, and the DNA-binding proteins in the nucleus.

The connection between an oncogene product and a normal growth factor was discovered with the help of a computer. An oncogene, sis, which had been identified in an ape sarcoma virus, was characterized by the DNA cloning and sequencing procedures that are now so common, and the DNA code was translated into the language of proteins. Two scientists in Uppsala, C. G. Heldin and Bengt Westermark, then fed this protein sequence into a computer that contained the sequences for all previously deciphered proteins. They asked the computer to search for any possible similarities. The search turned up one protein, or rather a part of a protein, that had previously been described and that had an almost identical structure. It was the growth factor PDGF (platelet-derived growth factor), normally released when blood platelets are disrupted during bleeding. The fibroblasts, cells of the connective tissue, carry a receptor that binds PDGF. This binding activates cell division, and the fibroblasts are stimulated to cover the wound. Of course, cells of the connective tissue also contain the PDGF gene, as do all other cells, but its activity is shut off.

The virus-borne v-sis gene that enters a fibroblast is continuously "switched on." During their evolution, viral genes have acquired the ability to compete successfully with the genes of the cell. They accomplish this by keeping their synthetic functions continuously "turned on" whenever the strategy of the virus demands. After its fortuitous capture by the virus, the cellular oncogene is "switched on" in a similar way. Now it was easy to understand why the ape viral v-sis gene caused a connective-tissue tumor! The infected fibroblast produces *its own* growth factor, which gives the cell an extra kick, starting a vicious circle.

Computer technology has also clarified the function of another type of oncogene. The sequence of erb-B, an oncogene isolated from a chicken leukemia virus, was fed into a computer. It turned out that its normal cellular counterpart, c-erb-B, is a receptor gene. It is responsible for the production of a component normally found on the epithelial cells of the skin and the mucous membranes. The product serves as the receptor for epidermal growth factor, EGF. Binding of EGF (the normal signal substance for cell division in epithelial cells) to its receptor triggers the cells to grow and divide.

But how can a receptor gene function as a cancer gene after being picked up by a virus? The answer to that question came when the normal c-erb-B was compared with the viral v-erb-B gene. The viral gene was shortened! It was lacking a small piece at the end, the most exterior part of the "antenna" that protrudes from the cell membrane and that normally binds the growth factor. In contrast with the previous case, this was not merely an illegitimate "switching on" of a normal gene; it was a change in its structure as well. The shortened "antenna" seemed to emit a continuous "green light" regardless of the presence or absence of the growth factor.

The remaining two classes of oncogenes are the largest and the most thoroughly studied. One is the ras group, so called since its first members were isolated from rat sarcoma viruses. Their protein products participate in the transfer of the signal from the receptor antenna to the inside of the cell. Tumor-producing ras proteins differ from their normal counterparts at only one single amino acid. The difference is due to a mutation that affects the protein's normal function in a way not yet understood. One can assume that some important signal is garbled or is otherwise wrong. The fourth group contains, among others, the important myc gene, originally isolated from a chicken leukemia virus. These genes code for a DNA-binding protein, and can contribute to tumor development when their normal regulation breaks down so that the gene is continuously switched on. The structure of the protein remains unchanged. Cells with a continuously expressed myc gene are prevented from reaching a normal resting stage in the cell cycle and therefore remain locked in a state of continuous cell division.

The information discussed so far was derived from occasional mistakes in the retroviral life cycle that led to the acquisition of cellular genes which were subsequently selected by scientists under very artificial conditions. There was no reason to believe that such viruses would ever survive and produce tumors in nature. They were pure laboratory artefacts, so to speak. How could it be determined, then, whether the oncogenes discovered by their occasional hitchhiking in viral vehicles played any role in the natural development of tumors? The answer to this question came from several other directions.

The Rous virus and similar acute tumor viruses were only the first of a number of windows that were opened for us recently, allowing new insights into the unknown world of oncogenes. Two other windows were opened by experiments that had nothing to do with viruses at all.

In the first of these experiments, the question was asked whether tumor DNA could convert normal cells into tumor cells. It was found that DNA from a human bladder cancer could indeed transform normal mouse connective-tissue cells into tumor cells! The fact that the DNA was derived from another species and from a different kind of tissue evidently didn't prevent it from functioning as a cancer gene in the mouse cell. Thanks to the species difference, a "handle" was now available on the integrated human DNA. Human and mouse DNA differ with respect to certain species-specific sequences, the so-called repetitive sequences. With the help of these repeats, it was possible to fish out the human DNA from the mouse cells and determine which human gene was responsible for the malignant transformation. It turned out to be an old acquaintance: the ras gene, previously isolated from rat sarcoma virus. Similar positive results were obtained with DNA from other human tumors, including lung and breast cancers and from certain leukemias. Even in these cases, it was often the ras gene that was responsible for the transformation. On some occasions other previously unknown oncogenes were identified. DNA from normal human tissues had no transforming effect.

It was assumed that there must be some important difference between the ras gene of transformed cells and the normal ras gene. DNA analysis revealed that the ras gene derived from the tumor contained exactly the same point mutation that had been found earlier in the gene from the rat sarcoma virus!

Still another window was opened by means of chromosome analysis. Together with Francis Wiener, George Manolov, and Lore Zech, we studied the chromosomes in certain mouse and human tumors derived from antibody-forming B lymphocytes.

These cells produce large amounts of immunoglobulins (Ig), the molecules with antibody function. There are three genes that take part in this process; one produces the antibody heavy chain and the other two produce the light chains. All three genes are located on different chromosomes.

Discriminating chromosome studies became possible through the new techniques developed by Torbjorn Caspersson and Lore Zech at the Karolinska Institute during the early 1970s. Their "banding" technique revolutionized all research and diagnosis by making it possible to recognize the individual chromosomes of all species, just as one recognizes the faces of acquaintances. B-cell tumors in humans, mice, and rats showed regular chromosomal changes, generated by the exchange (called "translocation") of certain segments between two different chromosomes.

Every tumor contained one of three characteristic translocations. They arose when the end piece of the human No. 8 chromosome or the mouse No. 15 chromosome was broken off at the same point each time and exchanged places with the most distant fragment of one of three different chromosomes that served as alternative recipients. Using the new chromosome techniques, we were astonished to find that the recipient chromosomes were those that contained the three Ig genes. We hypothesized that the misplaced fragment of the chromosome contained an oncogene situated exactly at the breakpoint, and that the translocation brought the oncogene into an active gene environment.

In an Ig-producing B cell, the Ig genes are constantly switched on, just as the silk gene is turned on in the salivary gland of the silkworm. If the normal function of the oncogene was to regulate cell division, in the manner of a traffic light, the translocated oncogene could malfunction, like a traffic light stuck on green.

In 1982, molecular biologists in the United States and in Australia showed that the hypothesis was correct. An oncogene was located exactly at the chromosome breakpoint, and it was the same oncogene in the B-cell tumors of mouse and man. Once again, the old acquaintances appeared in a new guise— the previously mentioned oncogene myc which had been isolated from a chicken leukemia virus. But in contrast to ras, the myc gene did not contain a mutation that could have explained its tumor-producing effect. The myc gene product binds directly to cell DNA and regulates cell division in ways that we still do not understand. The myc gene can get stuck and drive cell division relentlessly forward, either when it is moved from its normal position to a different chromosomal environment which is constantly active in the cell or when a normally non-transforming retrovirus happens to insert itself next to the myc gene. The translocation probably happens at a strategically important time in the life of the B lymphocytes: when they are programmed to move on to a resting phase after a period of active growth. The normal myc gene located in its correct place in the chromosome obeys this program, but a myc gene that has translocated to an unusual chromosome environment does not. It no longer "knows" that it is a myc gene, and it remains active along with the immunoglobulin gene to which it is attached. The cell therefore cannot enter its resting phase.

Translocations are rare accidents that can occur during cell division. The probability of such an accident is very small, but it increases with the number of cell divisions, just as the number of car accidents increases with increasing traffic. There is no special reason why just the myc-carrying chromosome should break and why this should happen in the immediate vicinity of the myc gene. The chromosomes that contain immunoglobulin (antibody) genes have more reason to break. They contain certain "hot spots," areas especially susceptible to breaks owing to a rearrangement of these genes during normal cell development. This process leads to the enormous diversity of antibodies. But the myc-containing chromosomes can break just above or below the gene at many different places, provided that the myc-protein coding regions are not damaged. This broad distribution of the breakpoints indicates that the "accidents" strike the chromosomes at random, and that the process of selection favors the malignant cell that is able to grow out as a tumor.

Burkitt's lymphoma is the human tumor most often associated with translocations of the myc gene. It particularly strikes children in the tropical rain-forest regions of Africa and New

Guinea (see chapter 8). During the past several years, the disease has also increasingly affected HIV-infected homosexual men between 20 and 40 years of age .

Besides the translocation, both the African and some of the AIDS-associated cases of Burkitt's lymphoma also carry the Epstein-Barr virus, or EBV (see chapter 10). This virus has a powerful ability to transform short-lived B lymphocytes into continuously growing (immortal) cell lines in tissue culture, but in a person with a normal immune system the cells are kept under control. After translocation of myc to an immunoglobulin gene, the cells grow into a malignant tumor.

EBV is found in 80–90 percent of all people in all countries. How can one explain the fact the Burkitt tumors preferentially strike African children and AIDS patients?

The "Burkitt's belt" in Africa and New Guinea corresponds to the distribution of hyperendemic malaria. This is an especially serious form of tropical malaria that appears in areas where the majority of the people are repeatedly infected with new malaria parasites at least once a month. Just as occurs with EBV, the malaria parasites can stimulate B lymphocytes to divide. They also weaken the "killer-cell response" against virus-infected B lymphocytes, which is mediated by the other main category of lymphocytes, the T cells.

Constant stimulation of cell division increases the risk of all genetic accidents, including myc translocations. The impairment of the T-cell response prevents the elimination of the cells carrying the translocation.

A similar situation can arise in homosexual men who suffer from chronic inflammation of their lymph nodes due to infection with HIV and other microorganisms. The latter may assume the B-cell-stimulating role of the malaria parasite while the AIDS virus overpowers the T-cell defense mechanism.

There is also another human malignancy, namely chronic myeloid leukemia (CML), in which oncogene activation by chromosome translocation plays a crucial role in the origin of the disease. The translocations in CML resemble the myc translocations in B-cell tumors, but they also differ in interesting ways.

In the 1960s, two scientists working in Philadelphia, Peter Nowell and David Hungerford, discovered that the smallest human chromosome, No. 22, was shortened in leukemia cells of this type. The small chromosome with the shortened small arm was called the Philadelphia or Ph 1 chromosome. Only one of two No. 22 chromosomes was of the Ph 1 type; the other was completely normal.

CML is a relatively mild form of leukemia. It may be viewed as a precursor to a truly malignant form of leukemia: the acute "blast crisis," which appears after three or more years. In some cases the Ph 1 chromosome is duplicated during the blast crisis, but at other times different chromosome changes are found. The Ph 1 chromosome is therefore linked to the development of this chronic and relatively mild leukemia.

Through the use of the chromosome-banding techniques, it was possible to search for the deleted piece of chromosome 22. In most cases, it was found attached to the end of chromosome 9. Thanks to oncogene research and to the development of molecular biology, it became possible a decade later to identify an oncogene, c-abl, at the extreme tip of chromosome 9. This oncogene had previously been isolated from a mouse leukemia virus. In humans, c-abl is located so far toward the end of chromosome 9 that it was not possible to determine by ordinary chromosome studies whether the c-abl oncogene had remained behind on chromosome 9 after the piece from chromosome 22 was attached there, or whether it wandered over to the remaining stump of 22, as would be expected in a usual "reciprocal translocation." The latter proved to be correct. The situation is similar to what was described earlier for the myc gene in B-lymphocytic tumors, but there is one important difference. The myc gene, activated by the translocation, produced a completely normal protein, but its transfer to a highly "activated" region of the chromosome resulted in its uncontrolled activity. In contrast, the Ph 1 translocation results in the production of an abnormal protein. The head of the c-abl gene had been chopped off, and the decapitated body was joined to a completely different gene that was also disrupted by the break on chromosome 22. Together, they produced an abnormal protein.

The translocations, "experiments of nature" as it were, have confirmed one important conclusion, even with the small amount of data from these two systems of "unnatural" virus-carried oncogenes. Clearly, oncogenes can contribute to the development of cancer either through an "illegal" activation of the normal gene or through a structural change, leading to an abnormal protein product.

The saga of the cancer genes ends here, for the time being. Like other great sagas that have accompanied us from ancient times, this tale has a beginning but no end. It will swell like a great river and appear in many different forms in many different areas. In contrast to the sagas of Orpheus and Prometheus, the Flood and the Tower of Babel, Wotan and Brünnhilde, Väinämöinen and Lemminkäinen, our epos is anchored—at first tenuously and vulnerably, later increasingly strongly—in an objectively verifiable reality that exists within us for both good and evil.

Epilogue

Bethesda, 1981. I have been invited to attend a symposium in honor of the founder of the oncogene concept, Dr. Robert Huebner, who is getting ready to retire from his position at the National Institutes of Health. My task is to summarize Huebner's impact on the development of our concepts in the field of cancer.

Before he became interested in the field of cancer during the period of great conceptual change in the 1950s, Huebner was already well known as a virologist and especially as an epidemiologist. He was very successful in tracking previously unknown viruses in our environment and in the world of our domestic animals. An indomitable will, remarkable energy, and enormous joy in his work—these were the most common epithets that his colleagues and co-workers used when speaking of him. They spoke highly of his instinct for what was essential and of his great confidence in his own intuition, which was often right on target in the field of infectious diseases that he knew so well. To be sure, there were occasional complaints about his lack of patience in listening to others. He was a man of great achievements and great mistakes. At the time of the great investment in virus-oriented cancer research during the Nixon administration, Heubner was one of the main actors on the scene, controlling hundreds of millions of dollars. He succeeded in setting an entire fleet of scientists into motion. They followed him into many new pursuits whose goals were decided by Huebner virtually alone in detached majesty. Nevertheless, he remained a popular, affable, and actually quite modest person at all times, a true American in the best sense of the word, available to all.

Huebner devoted his free time and private life to his bull-breeding ranch, which he managed, with his wife and nine children, without any additional help. The bulls were kept under model conditions and the offspring raised according to the latest and most modern concepts of cattle breeding. In parallel with his scientific work, Huebner was a regular contributor to cattle breeding journals, one of which honored him with a special issue. The cover was adorned with a picture of Dr. Huebner and his nine children and carried the title "The Givings of the Sire." I've often wondered if Huebner wasn't more proud of that picture than anything else.

Huebner's bull breeding became very successful, and sperm from his animals was sold all over the globe. Colleagues and friends often considered this a symbol of Huebner as a perpetual fount of ideas and as a grand entrepreneur, a fertility symbol in many respects. The publisher of the breeding journal put his finger on this.

But what difference does it make if the Freudian motive is more transparent with Huebner than with the rest of us? The issue is: What has he accomplished?

After the discovery by Gross, Huebner became one of the most important spokesmen for the cancer-virus theory and probably its most daring and unrestrained generalizer. He proselytized for Gross' concept that every form of cancer was caused by a virus. According to this theory, carcinogenic chemicals, radiation, or hormones would produce cancer only if they activated a latent virus. There was no such thing as cancer without a virus. The lack of positive evidence for the presence of virus genes was due to inadequate methods of detection, bad luck, or ineptitude.

It was against this background that Huebner formulated not only the concept of oncogenes but even the word. RNA tumor viruses were thought to contain two types of genes: virogenes and oncogenes. Virogenes were essential for the production of virus but not for tumor development; oncogenes were responsible for the transformation of normal cells into tumor cells. There was a certain amount of truth to this categorization, but two of Huebner's postulates were wrong. He thought that every RNA tumor virus, the acute ones as well as the chronic ones, contained an oncogene. The chronic viruses, of which the Gross and Bittner viruses are examples, were especially important parts of his reasoning. They were transmitted vertically, could remain latent, or could, according to theory, be activated by cancer-producing substances or radiation. But, as we already have seen, only the acutely transforming viruses, such as the Rous virus, contain oncogenes—not the chronic viruses. The viral origin of the oncogenes was the second mistake of the theory—they are derived from cells.

Real viral oncogenes do exist; however, ironically, they are not found in RNA tumor viruses, where Huebner searched for them, but rather in the other viral realm, among the DNA viruses. There are lessons in this story as well.

Huebner first became interested in DNA tumor viruses before he began to work in the RNA field. It was the polyoma virus in particular that awakened his interest. This mouse virus is a member of the same family that includes the rabbit wart virus described at the beginning of this chapter. But polyoma, discovered shortly after the Gross virus and largely responsible for the change of climate in tumor virology, causes at least a dozen different kinds of malignant tumors, provided that it is injected into mice less than 24 hours old whose mothers have not transferred any protective antibodies to them.

The polyoma virus was isolated from cell cultures, and experiments with the induction of tumors were carried out with laboratory mice. Good epidemiologist that he was, Huebner wanted to establish whether it could also could cause tumors in a natural environment. He laid mousetraps in the slums of Harlem and in the cornfields of Iowa to get representatives of both city and country mice. The trapped wild mice were tested for the presence of serum antibodies against polyoma virus and were found to be almost 100 percent positive. They were then housed in specially constructed cages in Huebner's laboratory—wild mice escape from their cages much more quickly than the docile laboratory strains. The investigators waited for the polyoma tumors, but none ever appeared. The mice grew old in radiant good health.

It is easy to understand this from today's perspective. As was described in more detail in chapter 10, most DNA tumor viruses commonly achieve a harmonious relationship with their natural host species. In the case of polyoma, the mothers transfer antibodies against the virus to their newborns, thereby protecting them from infection during the critical neonatal period. The immune system of the offspring has time to mature before this maternal antibody "umbrella" disappears. If the mice are infected later, their immune system, having been exposed to the selective pressure of the virus during the earlier evolution of the species, can easily recognize the polyoma-infected cells and reject them. But Huebner didn't know that. He concluded that polyoma could not be considered a tumor virus under natural conditions. He completely missed the important point that DNA tumor viruses contain their own oncogenes, corresponding exactly to the situation that he was searching for. Such viruses couldn't have survived in nature, however, if their host species hadn't succeeded in protecting themselves immunologically against the virally transformed cells. This was not particularly difficult, since the viral oncogenes were foreign to the species. Their protein products could easily be recognized as "foreign" by the host's immune system, in contrast to the RNA-virus-borne cellular oncogenes, which produced indigenous proteins of the host species and therefore were are recognized as "self."

After his experiments with the polyoma virus, Huebner lost all interest in DNA tumor viruses. He switched over to the RNA tumor viruses, where he hoped to find authentic viral oncogenes. But he searched for them in the wrong place: in the chronically acting viruses, which had no oncogenes. They could have been found in the acutely acting viruses (although they were of cellular origin, as we have already seen above), but

Huebner wasn't at all interested in them, since he understood quite correctly that they were artefacts of the laboratory and could not produce tumors in nature.

Nature will not be coerced. Huebner was correct in principle, but he chose the wrong system to prove it. It fell to other scientists to find the correct systems and to arrive at the correct interpretations. Huebner's accurate intuition as an epidemiologist of infectious diseases led him astray when he went over to cancer, a cell disease. Viruses can certainly cause cancer, but the cell is always the focus. The relationship of a virus to a cancer cell is indirect and complicated. The DNA tumor viruses have their own genes that can transform normal cells into cancer cells, but these genes produce "foreign" proteins, which present a relatively simple target for the immune system. RNA tumor viruses cannot transform on their own; they do so only accidentally when they happen either to insert themselves next to (and thereby activate) a cell-growth-regulatory gene or to capture such a gene and transfer it to a susceptible recipient cell.

This father of nine, the bull breeder and unending source of inspiration, Huebner the virus hunter, the man who overflowed with ideas, research projects, grants, contracts, and laboratory constructions like no one before him, he who tirelessly stimulated and prevailed upon others but who never listened to anyone except his own God, has come to the end of his journey in the rock-strewn field of cancer. His God has toyed with him and played a practical joke on him. Like the patriarch Jacob, Huebner has had his triumphs and defeats revealed as illusory. He was right when he thought he was wrong and wrong when he thought he was right. There in the apparently secure nest of the viral genes, wrapped neatly in their virus coat, sat the cuckoo egg—the oncogenes—the most important growth-regulatory genes of the cell, offering their services to happily astonished cell biologists and molecular biologists while their discoverers, the classical virologists, faded from their stormy and often heroic years into the history of cancer research.

La Condition Humaine

12

Blind Will and Selfish DNA

Cambodia, 1966. Deep is the jungle around Angkor Wat. The monumental temples of the Khmers have fused with the vegetation. The giant roots of the trees have crept in among the stones during centuries of slumbering oblivion, advancing along the path of least resistance, implacable, inexorable. Now they form an inseparable part of the god-kings' eternal temples. The inextinguishable yearning of the sons to build temples grander than those of their fathers and conquer ever vaster territories led to overdevelopment, leaving decadence, foreign invaders, destruction, and death in its trail. Men vanished and tropical vegetation took their place. Only centuries later would they return. The amazed French archaeologists opened the way, followed by the Buddhist monks.

Underneath the main cupola of the largest temple the Golden Buddha sits impassively, looking out at the passing centuries with an unchanging, gentle smile. In the arch above his head live tens of thousands of bats. At night, they fly out above the Buddha's head. They are steered by the sonar navigation instrument lying somewhere behind their blind, foxlike faces. They move about quickly and silently with unbelievable precision. But in the daytime, they hang from the ceiling like small, insignificant bags. When they die, they fall and remain lying around the Buddha. Huge tropical insects swarm all over them, They look like the sacred scarab beetles of the Pharaohs, but with golden wings. Visitors are warned of their poison. They quickly dispose of all the soft parts; they convert the organic matter to produce new eggs, larvae, multifaceted eyes, new wings. We keep an appropriate distance, but we notice that the

insects do not frighten Buddha's devoted servants who come
with their offerings. They bow, recite unintelligible prayers,
hang colored ribbons and flowers on the boards around the
Buddha, or stuff small rolls of paper at his feet. But the Buddha
doesn't pay any attention to them. He looks out through the
opening of the arch with his gentle smile and views the perpetual change of all living things with equanimity. Centuries come
and go, empires are built and overthrown, kindly moods of
tolerance and friendship alternate with bloody violence, war,
and ideologically rationalized mass murder. The bats, the
insects, the stones, the trees, and the people with their wretched
and pitiful prayers all continue their eternal cycle. The seasons
come and go. But the hand of Buddha never trembles.

Jerusalem, 1985. I'm standing by the window in my study at the
Hadassah Medical School, dictating a paper on the evolution of
tumor cells. Jerusalem's famous yellow building stones shine in
the dazzling sunlight, students are moving about in the court-
yard, and a pleasant breeze drifts in through the open window.
Sounds, subdued and unobtrusive, come from the laboratory
across the corridor. I find myself in a pleasant intermediate
state between dead silence and the intrusion of the spoken
word. I note with satisfaction that my opportunities for concen-
trated work are as good as they can possibly be.

Suddenly, a big fly comes in through the open window—an
annoying, grayish-black thing making a very unpleasant sound.
I try to ignore it. There's a very high ceiling, so why shouldn't
there be room for us both? Why should I disturb my peace of
mind to go on a hunt? I try to work but can't. The crystal of my
concentration has turned into foggy nervous frustration. A
stone has been thrown into the pond, and the lovely surface
reflection has broken up into thousands of grotesque caricatures, quivering in all directions.

I really have to stop this. I go after the fly with a rapidly
growing rage. There's only one thought in my head: to kill the
intruder, ruthlessly and quickly. After a lengthy hunt I finally
get the fly, and it falls on the windowsill. Finally I can get back
to work!

I go back to the same spot by the window and try to put the pieces of my thoughts back together. But they're either hiding somewhere, damaged by all the commotion, or just sitting there, sulking. While I wait, my eyes are drawn to the gray-black corpse on the windowsill. It really is dead—good shot! Dead now and for eternity, a corpse like I will be one day. But . . . what's happening? Is it moving? Didn't I kill it? My rage, which had already fallen asleep and hidden beneath the covers, begins to stir again.

But what kind of motion is that? A fly can't move like that by its own free will. It's the back part of the corpse that is moving, in a somewhat passive way, like a small pebble that someone is toying with. What's happening now? Something white is sticking out of the hind part of the fly—a small worm, wriggling back and forth. I'm suddenly so overwhelmed by a terrible disgust that I almost throw up. Could that big, beautiful fly have been harboring a parasite? No, that can't be! Now, suddenly, many white worm heads stick out—or are they tails? Five, ten, twenty— they creep out, moving quickly all over the dead fly, searching desperately, but for what? It's obvious: they're searching for their mother! The fly must have been a female with fully developed eggs, ready to hatch. Some incomprehensible signal must have told the larvae that their mother was dead, and now they've been ejected just like a pilot from a burning airplane. Now they're all out, something like fifty of them, in a wild, searching protest. They fan out in a wide circle on the windowsill, searching perhaps for a foster mother or a new, dark, protective abdomen—or possibly some kind of compensation for the nutritious security in which they had been lying a little while ago, packaged with an efficiency that makes the world's most efficiently designed sardine can seem an absurd parody.

My disgust vanishes and is replaced by amazement and wonder. The fly is one of the marvels of evolution, just as I am. The larvae, now about to end their wanderings and to slip into the peaceful submission of death on the sun-baked windowsill, have just as long a history of evolution behind them as I do. The only difference is that theirs has taken a different course. Why does my DNA have greater rights that theirs? We utilize the same genetic code language—we are brothers! Many of my most important genes are to be found in them as well. They breathe; they burn sugar for energy; their cells begin and end their divisions, just as mine do. They certainly don't have much of a brain, but they do have millions of other skills that I don't have. The entire accumulated history of human art is like child's play compared with the unbelievable masterpiece of the multifaceted eye. The aeronautical precision and quick flexibility of the wings compares to the most sophisticated computer-navigated fighter plane as a Bach fugue played by a master to a primitive electronic imitation.

The larvae die, and my admiration is mixed with humility and guilt. But, at the same time and on a different level, something else happens. What sort of association is emerging, or could it be a memory of something? I feel the mood of it long before I know what it is. It is a relic from my adolescence, a long-forgotten "Aha," an experience of discovery of some kind, a sort of "of course, that's the way it is, it fits exactly right!" I feel the beat of wings that blend the joy of insight with the incurable sadness of the inevitable. Where do I find the words that tell me when, how, and why I had experienced this mood? I feel that they're coming, forcing their way out, but they are pushing a large mass of false gravel in front of them, generating a lot of meaningless noise. Relax; think of something else.

I walk home through the grass, among the flowers and past playing children. Down below I can see Ein Karem, John the Baptist's village, with its many churches, sublimated symbols of mankind's longing but also of its watertight compartments, the many religions that appear more ornate and more implacable in Jerusalem than anywhere else in the world. And suddenly the words appear there in front of me—in German. *Die Welt als Wille und Vorstellung* (*The World as Will and Idea*), Schopenhauer's most important work of 1818 and an unforgettable experience of my late adolescence. But of course! The will of the fly and its offspring has come into conflict with mine. She was destroyed— not I, not just yet, not at the moment. But our DNA lives on, to continue its struggle to dominate the world, equally blind, hopeless and inexorable, with the same purpose or lack of purpose, depending on how one wishes to interpret it. At the same time, I, but not the fly, have the ability to formulate certain concepts about both of us and about our world. This capability is a product of my brain, evolved as an instrument of battle in the same category as the snake's venom, the wolf's fangs, and the wildcat's claws. But the structure of the brain came to include even the capacity for speech, with its powers of abstraction, ideation, the conception of time, and, sometimes, the capacity for detached contemplation. On rare occasions, it extends from the world of the will to the world as an idea, freed from time and space and even from causality, in the terms of Schopenhauer.

Is it really so? Yes, possibly, in some respects. But the world as an idea can never be completely free of a certain mood. The basic tone can be sorrowful or elevating, pessimistic as in the case of Schopenhauer or filled with a certain optimism, as in the writings of most of the other nineteenth-century philosophers. But must an acceptance of Schopenhauer's pessimism make us sad? What should I mourn—my impending death as an individual, a day, a year, or a decade hence, ephemeral and meaningless periods of time under the skies of eternity? Or should I lament the blindness of will, whose trap keeps us imprisoned just like the fly larvae strapped into their ejection seat? No, "mourning becomes Electra." Let others mourn—I have no time for that. The ideas that were born from contemplating the blind will of the larvae are leading me toward other landscapes.

Selfish DNA

Schopenhauer knew nothing about DNA. Nevertheless, there are strange parallels between his "world as will," the urge that drives all living creatures, and modern discussions on the "selfishness" of DNA.[1] Long before Darwin, Schopenhauer wrote:

. . . everywhere in nature we see strife, conflict, and alternation of victory. . . . Every grade of the objectification of will fights for the matter, the space, and the time of the others. . . . This universal conflict becomes most distinctly visible in the animal kingdom. For animals have the whole of the vegetable kingdom for their food, and even within the animal kingdom every beast is the prey and the food of another; that is, the matter in which its idea expresses itself must yield itself to the expression of another idea, for each animal can maintain its existence only by the constant destruction of some other. Thus, the will to live everywhere preys upon itself, and in different forms is its own nourishment, till finally the human race . . . because it subdues all the others, regards nature as a manufactory for its use. Yet even the human race . . . reveals in itself with most terrible distinctness this conflict, this variance with itself of the will, and we find *homo homini lupus.*

(*So sehen wir in der Natur überall Streit, Kampf und Wechsel des Sieges. . . . Jede Stufe der Objectivation des Willens macht der andern die Materie, den Raum, die Zeit streitig. . . . Die deutlichste Sichtbarkeit erreicht dieser allgemeine Kampf in der Tierwelt, welche die Pflanzenwelt zu ihrer Nahrung hat, und in welcher selbst wieder jedes Tier die Beute und Nahrung eines andern wird, d.h. die Materie, in welcher seine Idee sich darstellte, zur Darstellung einer anderen abtreten muss, indem jedes Tier sein Dasein nur durch die beständige Aufhebung eines fremden erhalten kann; so dass der Wille zum Leben durchgängig an sich selber zehrt und in verschiedenen Gestalten seine eigene Nahrung ist, bis zuletzt das Menschengeschlecht, weil es alle andern überwältigt, die Natur für ein Fabrikat zu seinem Gebrauch ansieht, das selbe Geschlecht jedoch auch . . . in sich selbst jenem Kampf, jene Selbstenzweiung des Willens zur furchtbarsten Deutlichkeit offenbart, und* homo homini lupus *wird.*[2])

Schopenhauer sees the will as the most powerful driving force in all species. It constitutes the basis for the entire biological world and can be understood as the will for life. According to Schopenhauer, the will for life has no real purpose whatsoever. It aims neither to find pleasure in life nor to understand its phenomena. It is simply a blind urge that must perpetuate itself. It is a *Ding an sich,* a thing unto itself, the only thing that exists, the innermost core of all things. When the will takes form in living organisms, it is split into an enormous number of specimens in each species, according to the "principle of individuation." Despite this split, the will is expressed with undiminished force in every one of the billions of organisms that are produced. A casual observer is bewildered by this process and often misses the basic principle of individualization, which states that

the will itself, the thing-in-itself, is without ground . . . for it lies outside the province of the principle of sufficient reason. Therefore, every man has permanent aims and motives by which he guides his conduct, and he can always give an account of his particular actions; but if he were asked why he wills at all, or why in general he wills to exist, he would have no answer, and the question would indeed seem meaning-less; and this would be just the expression of his consciousness that he himself is nothing but will, whose willing stands by itself and requires more particular determination by motives only in its acts at each point of time. In fact, freedom from all aim, from all limits, belongs to the nature of the will, which is endless striving.

(*der Wille selbst, das Ding an sich, ist grundlos . . . daher hat auch jeder Mensch beständig Zwecke und Motive, nach denen er sein Handeln leitet, und weiss von seinem einzelnen Tun allezeit Rechenschaft zu geben: aber wenn man ihn fragte warum er überhaupt will oder warum er überhaupt da sein will, so würde er keine Antwort haben, vielmehr würde ihm die Frage ungereimt erscheinen; und hierin eben spräche sich eigentlich das Bewusstsein aus, dass er selbst nichts, als Wille ist, dessen Wollen überhaupt sich also von selbst versteht und nur in seinen einzelnen Akten, für jeden Zeitpunkt, der nähern Bestimmung durch Motive bedarf. In der Tat gehört Abwesenheit alles Zieles, aller Gränzen, zum Wesen des Willens an sich, der ein endloses Streben ist.*[3])*

"Blood is a wondrous juice," Mephistopheles tells Faust. But everything found in the blood is a product of DNA. The instructions for all life on this planet are written in the increasingly well-understood code language of DNA. This most ingenious of all molecules combines stability with flexibility, the greatest precision with irresponsible carelessness, conservatism with capricious changeability. Its alphabet has only four basic letters, but the combination of three letters at a time produces a real "hieroglyph," a symbol that represents a specific product: an amino acid. This allows 64 different combinations of words—many more than are needed, since only 20 amino acids exist. The code is therefore "degenerate" in the language of information theory—several different hieroglyphs can have the same meaning. The code is universal: the life program for all living things— plants, animals, humans, bacteria, fungi— is written in the same language. It is this alphabet that spells out the venom of the snake and the silk of the silkworm. It carries the blueprint for brains that can think Plato's thoughts, write Mozart's music, and blow open safes. The Pharaohs' futile yearning for an eternal life and their dreams about the Sun God and the Father are all based in the blueprint of DNA, as is the power of the giant roots that force the walls of the Khmer temples apart. It is the instructions of DNA that drive bats on their nightly hunt and determine their sonar code, and it is the same DNA that gilds the wings of tropical insects. It prompts humans to dream about the eternal Buddha who will liberate them from their endless slavery.

But DNA isn't merely a code. The sum total of all the texts written in its language is far richer than everything else written with our human scripts, whether they be primitive runic inscriptions, complicated hieroglyphs, or the somewhat more realistic Chinese pictograms. But in contrast to written symbols, which must be read by someone before their instructions can become reality, the DNA molecule has two unique characteristics: it makes copies of itself and it drives the production of the machinery needed to carry out its instructions. In other words, it has the dual capacity for self-replication and for function.

The capacity for replication is based on the structure of the DNA molecule, the famous double helix of Watson and Crick. The molecule consists of two intertwined strands. One contains the correct code text as it is to be read, and the other is its mirror image. The relationship between the two strands is more or less the same as the relationship between a photograph and its negative: the negative strand is used to produce a new positive strand, and vice versa.

The photographic copying process requires a series of external devices. Those tools are very modest in comparison with the enormous apparatus needed to to produce a new "mirror image" strand of DNA. But in contrast to the example of photographic copying, the original strand of DNA contains all the information required to produce such an apparatus. But this is not all. The text even directs its own proofreading. The cell contains a whole forest of proteins specially made for this purpose. Some, called "repair enzymes," are prepared to step in if something goes wrong. Some survey the newly copied strand and correct mistakes in the letters or the words. Some of them can cut out long pieces and sew in a new and improved copy. Others have something more like a bookbinding function; they glue together pieces of DNA that, for technical reasons, are produced in separate sections. Every one of these proteins is manufactured from its own blueprint found in the DNA text. In other words, the DNA carries within itself all the information needed for its flawless duplication, together with the negative that has to be copied. And all of these are still only a few of the many facets of DNA.

The nucleus of every cell in every organism contains all the information that each cell will need during the organism's entire life. But only the cornea of the eye is transparent, only cells of the retina have an apparatus that can respond to light, only kidney cells can filter urine from the blood, only muscle cells can contract to bend an arm or pump blood to the heart, only the red blood corpuscles can transport oxygen. And yet all cells have the same DNA, a faithful reproduction of the single sperm and egg nuclei united at the moment of conception. Many billions of cells are produced from the single fertilized egg cell before we leave our earthly life, but even the very last cell we make will be a true replica of the original paternal and maternal chromosomes as they first met in the fertilized egg.

The functional diversification of cells is due to the fact that only about 2 percent or even fewer of all the genes are functioning in each cell. Different "specialized cells" use different parts of the genetic repertoire. But how does the cell know which page to open in the instruction manual?

If the book of instructions is a world of DNA, its product, the cell, is a world of specialized proteins that carry out the instructions. The relationship between the language of proteins and the language of DNA is like the relationship between an actual lamp and the individual letters *l-a-m-p*—or, more accurate, a very detailed description of all the lamp's component parts. Every protein has its corresponding DNA instruction, and every set of instructions has a beginning and an end. A specially adapted protein—an enzyme—reads the instruction. But it has access only to a very limited part of the DNA, areas that are like small open windows on the facade of a big building. Different types of cells keep different sets of their windows open. The time and place for the opening and closing of all the windows are regulated during embryonic life by a meticulously precise program. The enzyme responsible for reading the instruction can also write while it reads. It makes a small strip of tape, a kind of telex message or quote. It is written in the language of RNA, which differs from the original DNA language only in some small details—a sort of dialect, so to speak. These strips are called "messenger RNA," a beautiful name with strong mythological associations. But unlike Hermes, with his messenger's rod and the small wings on his heels, messenger RNA is far from being completed at this stage. It must first be transferred from the large "reading room" of the cell nucleus to the cell's cytoplasm with its protein factories. It passes several barriers on its way and is extensively trimmed. The message contains large and apparently unnecessary pieces that must be cut out. Why are they there at all? Are they living fossils from some ancient time, or do they have some as yet unknown function? No one knows, but the extensive removal of large bits of the presumably vital code text puzzled many scientists and raised many eyebrows long before there was any serious discussion of the "selfishness" of DNA. I will return to this point.

The trimmed messenger is also provided with many different kinds of small labels that are required to guide it through the large and complicated structures of the cell cytoplasm. It has to reach the protein-synthesizing factories, the ribosomes. This is where the instruction is translated into the final product. For every three-letter hieroglyph or "codon" there is a mirror-like "anticodon," carried by a molecule with a cloverleaf-like structure. At the other end, this molecule carries the corresponding correct amino acid. The whole mechanism resembles one of those marvelous Chinese typewriters where the pictogram symbols are picked out by hand, one by one; however, instead of stamping out the same symbols, protein translation generates the real product: if the head of the code word has a symbol for "lamp," then that very same little lamp is found already attached and shining on the tail. Each and every one of the twenty vitally important amino acids has its own translation molecule, a kind of "miniature Rosetta stone," that can crack the code. The code words are translated, one by one, and the protein chain is extended quickly and smoothly, apparently quite miraculously.

The whole process gives the impression of purposeful precision, extreme economy, and maximal effectiveness. At first glance, it is as if the architect of the evolutionary process, the mechanism of natural selection, has produced a magnificent piece of engineering machinery. But as François Jacob points out in his brilliant essay "Evolution and Tinkering,"[4] this analogy is completely inappropriate. An engineer works only from a prepared and specific plan, and he has both the material and the machinery at his disposal. The well-engineered product is as perfect as the technology of the time permits. But the work of evolution is far from perfect. The biological world is characterized by an almost unbelievable wastefulness, from individuals and their reproductive or germ cells to the total number of species. Simpson has estimated that the number of existing biological species is approximately 1 million. The number of extinct species is about 500 million.[5]

Jacob compares natural selection to a gifted amateur, a kind of tinkerer who does not know what he wants to produce but who uses all kinds of junk to construct some kind of functional machine. He is a master of improvisation. His material has no specific and clearly predefined function—it can be used for a number of different purposes. The innumerable tinkerers who toil in nature's laboratories have virtually unlimited time at their disposal. They modify their products constantly, taking a bit off here and adding a little there, until the finished product meets the requirements. They work on many different objects, and they often arrive at completely different solutions to the same problem. Jacob cites as an example the enormous variety of light receptors in the biological world. Superficially, nothing resembles the human eye as closely as the eye of an octopus. But in reality, these two structures represent the end results of different and completely independent evolutionary processes. They are built differently, and their functioning is based on diametrically opposite principles.

Jacob uses the human brain as the ultimate example of the tinkerer's achievements—ultimate in a double sense. The youngest part of the brain in evolutionary terms—the neocortex, which controls our intellectual and cognitive abilities—has simply been added to the earlier structure that controls both the emotional and the visceral activities, without any coordination or any clear hierarchical relationship between the two. Jacob compares this to an old horse carriage which suddenly has been equipped with a jet engine. He adds: "It is not surprising in either case that accidents, difficulties, and conflicts can occur."

What one of the century's most prominent molecular biologists sees as a horse carriage with a jet engine was for Schopenhauer the world as will and idea. He saw the relationship between the two in the following way:

Understanding, as a rule, is always subordinate to the role of the will and has obviously come into existence for exactly this purpose. It has arisen from the will exactly as a head springs out from the body. The subordinate role of understanding to will in the animal world can never be elevated. In the case of man, such an elevation takes place, but only as an exception.

(*Dem Dienste des Willens bleibt nun die Erkenntnis in der Regel immer unterworfen, wie sie ja zu diesem Dienste hervorgegangen, ja dem Willen gleichsam so entsprossen ist, wie der Kopf dem Rumpf. Bei den Tieren ist diese Dienstbarkeit der Erkenntnis unter dem Willen gar nie aufzuheben. Bei dem Menschen tritt solche Aufhebung nur als Ausnahme ein.*[6])

"There is nothing new under the sun" (*"Ein kol hadash tahat hashemesh"*), said Qoheleth, the son of King David, known as the Preacher.

Or is there?

Yes, there is. Its name is DNA.

DNA as the Ultimate Parasite

Yes, this is exactly what Lesley Orgel and Francis Crick called "selfish DNA."[7] Can we believe our eyes? Is this the same DNA and the same Crick? The magnificent DNA that drives all living things on this planet, whose double helix Crick himself discovered with James Watson? Is Crick no longer proud of his child, the Lord and Father of us all? Of course he is!

Once, when he was introduced to a group of physicians as "Doctor Francis Crick, who has a Nobel Prize," he quickly corrected the chairman: "No, you're wrong. I have *the* Nobel Prize." Yes, of course: if anyone has it, Crick does. His is the prize of the century.

But what is the meaning of "selfish DNA: the ultimate parasite"? The concept was formulated with the same special mixture of mild but challenging irony and deep seriousness, the same deliberately blurred and therefore highly provocative boundary between sarcasm and earnestness, that has always characterized Crick and his circle of friends. It stems from the unexpected discovery that an indeterminate but certainly considerable part of DNA is "junk" without any function. The junk has arisen in a variety of ways. A minor part consists of "dead genes" or "pseudogenes." These are genes that have "rusted away"; they are like weed-covered tracks on which no train has traveled for a very long time. Some of these pseudogenes have ended up in the wrong place because of some unusual accident during cell division, or after having "jumped" from one chromosome to another as genes occasionally do. If a gene lands in an inactive part of the chromosome, it may cease to function. A non-functional gene "rusts" through mutations that introduce stop signs or nonsense letters into a previously meaningful text. Functional genes are also exposed to the same mutational risk, but their "railroad track" of instructions is kept clean through natural selection, which quickly eliminates the non-functional mutants if they decrease the organism's fitness.

But the majority of the junk DNA does not consist of eroded genes. Its language, or rather the sequence of its letters, is completely unintelligible nonsense. It is more like a series of symbols pecked out on a typewriter by a small child who has not yet learned to read or even by a trained monkey. Letters and punctuation marks follow one another in a real jumble. Often it is not only a question of single bits of nonsense; *most* of the junk DNA consists of nonsense. An immersion into this microcosm is comparable to the archaeological digs in Jerusalem. Within a very limited area, archaeologists find mostly gravel and junk; but they can occasionally hit upon Byzantine mosaics, the wall of a Crusader's stronghold, a commercial street from the Roman era, a magnificent villa of Herod's kingdom, fragments of columns from the First Temple, or a Stone Age fireplace. Different cultures have been layered on top of one another, and no one has ever worried about removing one stratum before building the next one right on top of it. But the known and even the unknown heritage of humans is no more than a wisp of a thin veil when compared against the history of

DNA. The apparently useless structures indicate a past of far greater depth. Some of them might possibly play a mechanical role, like the steel skeleton of a concrete building or the solid rock under a village. But large regions of DNA are like surrealistic film landscapes, like the kingdom of Hades in Cocteau's *Orphée* or the Zone in Tarkovsky's *Stalker*: destroyed houses, demolished tanks, remains of animals long since dead, petrified human remains, witnesses of ancient weekdays, holidays, and wars. But, in contrast to old ruins, the useless parts of DNA are just as alive as their more useful counterparts: they multiply during cell division with exactly the same precision, for all eternity. This is the landscape of *selfish* DNA. Its only goal is to continue a never-ending process of "xeroxing"—a goal limited only by conditions of the external world, never from within.

The great drama of the biological world is only a small oasis in the gigantic and barren moonscape of selfish DNA. We see the final compromise between the blind will of DNA to reproduce itself and the unyielding pressure of selection. We would rather direct our view toward the beautiful scenery and turn our backs on the more alarming surroundings. But they belong to the same world—they *are* the same world. Even the most unbelievably perfect adaptations—the navigation of migrating birds and whales, the biochemical precision of unicellular organisms adapted to the conditions of light and temperature at the depths where they find themselves, the use of polarized sunlight for the completely instinctive dance by which bees signal the location of nectar, the ability of the chameleon to change its coloration depending on the background—are based on "chance and necessity" (to use Jacques Monod's expression)—the interplay of random events with unyielding natural law: genetic variation, the basis of natural selection. The law is written in the language of DNA. From the vantage point of DNA, the whole biological world may be seen as an instrument serving its aimless reproduction. Regarded in this light, DNA may indeed appear as the ultimate parasite.

La Condition Humaine

No s megvagyok már. Lettem ami lettem. Az Isten sem változtat rajta többe.
Here I am, I have become what I have become. Even God cannot
change that any longer.
Mihály Babits, *Nel mezzo . . .*

Our own species—a late and apparently still somewhat drowsy
product of evolution; a youngster that cannot be more than 2
or 3 million years old—is sitting at the top of the pyramid of the
biological world, a pyramid some 500 times older. Our unique
instrument, our large brain, still an almost incomprehensible
magic box, has given us this preeminent position. It has created
in us the first living creature capable of contemplating itself and
the world, the first creature that knows it must die. The world
of the will has opened the way for the world of ideas. The latter
existed at first only as a subjective experience, but it is now open
to objective verification at a level that would have been inconceivable during Schopenhauer's time. Would he have accepted
evidence of an objective reality that lies outside of our own
mind today? It is possible, but far from certain, perhaps not
even probable. But he would certainly maintain that blind will
continues to operate with undiminished strength. It is, in his
own words, "the inmost nature, the kernel, of every particular
thing and also of the whole. It appears in every blind force of
nature and also in the preconsidered actions of man; and the
great difference between these two is merely in the degree of
the manifestation, not in the nature of what manifests itself."
(*das Innerste, der Kern jedes Einzelnen und eben so des Ganzen: er
erscheint in jeder blindwirkenden Naturkraft: er erscheint auch im
überlegten Handeln des Menschen; welcher Beiden grosse Verschieden-
heit doch nur den Grad des Erscheinens, nicht das Wesen des Er-
scheinenden trifft.*[8])
 In comparison with the constant activity of the will, the
development of ideas is a very rare event, according to Schopenhauer:

The transition, which we have referred to as possible, but yet to be
regarded as only exceptional, from the common knowledge of particular things to the knowledge of the Idea, takes place suddenly; for

knowledge breaks free from the service of the will, by the subject ceasing to be merely individual, and thus becoming the pure will-less subject of knowledge. . . . If raised by the power of the mind, a man relinquishes the common way of looking at things, gives up tracing, under the guidance of the forms of the principle of sufficient reason, their relations to each other, the final goal of which is always a relation to his will; if he thus ceases to consider the where, the when, the why, and the whither of things, and looks simply and solely at the *what;* . . . inasmuch as he *loses* himself in this object (to use a pregnant German idiom), i.e., forgets even his individuality, his will, and only continues to exist as the pure subject, the clear mirror of the object, so that it is as if the object alone were there, . . . then that which is so known is no longer the particular thing as such; but it is the *Idea,* the eternal form, . . . and, therefore, he who is sunk in this perception is no longer individual, . . . but he is *pure,* will-less, painless, timeless subject of knowledge.

(Der . . . mögliche, aber nur als Ausnahme zu betrachtende Übergang von der gemeinen Erkentnnis einzelner Dinge zur Erkentnnis der Idee geschieht plötzlich, indem die Erkenntnis sich vom Dienste des Willens losreisst, eben dadurch das Subjekt aufhört ein bloss individuelles zu sein und jetzt reines, willenloses Subjekt der Erkenntnis ist. . . . Wenn man, durch die Kraft des Geistes gehoben, die gewöhnliche Betrachtungsart der Dinge fahren lässt, aufhört, nur ihren Relationen zu einander, deren letztes Ziel immer die Relation zum eigenen Willen ist . . . nachzugehen . . . also nicht mehr das Wo, das Wann, das Warum und das Wozu an den Dingen betrachtet; sondern einzig und allein das Was . . . indem man, nach einer sinnvollen deutschen Redensart, sich gänzlich in diesen Gegenstand verliert, d.h. eben sein Individuum, seinen Willen, vergisst und nur noch als reines Subjekt, als klarer Spiegel des Objekt bestehend bleibt . . . dann ist, was also erkannt wird, nicht mehr das einzelne Ding als solches; sondern es ist die Idee, *die ewige Form . . . und eben dadurch ist zugleich der in dieser Anschauung Begriffene nicht mehr Individuum . . . sondern er ist reines, willenloses, schmerzloses, zeitloses Subjekt der Erkenntnis.* [9]*)*

Timeless? Yes, sometimes. Without will (*willenlos*)? Possibly, but rarely. But painless? Isn't insight into the accidental nature of our existence, the exchangeability and absolute mortality of the self, one of the first concepts our brain encounters? Doesn't it lead inevitably to a head-on collision between Jacob's jet-powered horse carriage and Schopenhauer's world of will and idea?

Psychologists say that death is an emotionally unacceptable fact. None of us can grasp the idea that an endless time is to come during which we will not exist. The anguish aroused by this thought is in strange contrast to the complete equanimity

that characterizes our outlook on the endless time that preceded our existence. Yet we all know that we are going to die.
How do we solve this problem? Suppression is the most common reaction. We think, talk, plan, and act as if we had eternal
life. No longer can any of us behave like the Khmers or the
Egyptian god-kings and deny his own individual death, with all
the practical inferences that follow (including the agonizing
death of hundreds of thousands of slaves whose forced labor
built monuments to serve their rulers during their imagined
eternal life). But we are precisely the same biological creatures
as those god-kings. Our brain functions in the same way. Our
ever-anguished souls construct their own small cathedrals until
the moment of truth comes. Meanwhile the will, the selfish
DNA, carries on with its blind game. Periods of perilous calm
and security alternate with war and annihilation. Peace movements assert themselves now and then, fighting an uphill
struggle against the arms race. Technology, a product of our
brain, seriously threatens all life on the planet for the very first
time.

George Snell, father of transplantation immunology and winner of the 1980 Nobel Prize for medicine, comes from an old
New England family with strong puritanical traditions. More
than thirty years ago, he told me that after his retirement he
planned to devote himself to a problem far from his own area
of work that had concerned him all his life: Does the Christian
ethic give man a biological advantage after he has achieved a
certain degree of social development? He hoped that the
answer would be positive, but he was far from certain that such
an answer would be the outcome.

I recently met George Snell again. He was 82 years old, and
he had been working on this problem for a decade. I asked him
anxiously how much progress he had made. He said that he
didn't yet know with any assurance what the answer would be.
He still hoped that it would be positive, but much work remained to be done.

Another Nobel laureate, Carleton Gajdusek, who discovered
the cause of the disease kuru, has explored many areas of New
Guinea where no white man had ever set foot before him and

where people continued to live in Stone Age cultures. He gave a more definite answer to Snell's question: The polish of civilization provides the thinnest possible veneer over our primitive urges. It is torn apart like a fragile spider web under conditions of stress, and often even without any apparent stress. According to Gajdusek, no ethical rules can have a serious impact on this basic fact, even over long periods of time. Nor are there any profound differences between Stone Age societies and highly civilized people in this respect.

Since I was in Angkor, one and a half million Cambodians have been murdered by an ideologically justified genocide, the first of its kind. Foreign invaders again rule over this gentle, friendly people. The village of Siam Reap is said to have been destroyed. The temples of the god-kings are said to be still standing. It may be presumed that the jungle vegetation has now become even thicker and darker, but the statue of Buddha looks out over the landscape with the same gentle smile. Another century is passing by in front of him, on its way to well-deserved oblivion. But his raised hand never trembles.

13

Eternel Printemps
(A Kaleidoscope of Sexuality)

Sento un certo non so che, che mi pizzica e diletta, dimmi tu, che cosa egli è?
I feel something, I don't know what; it tickles and delights me—tell me, what is it?
the page to the girl, in Claudio Monteverdi's *Coronation of Poppea*

Budapest, 1939. I've just visited a friend, about my own age, and traded some stamps with him. A colonial stamp that I had never seen before carried the strange place name "Ruanda." I look it up in the big encyclopedia. While turning the pages I come upon a full-page picture of Rodin's statue "Eternel Printemps."[1] I stare at it, completely fascinated. The stamp collection, my friends, school, the everyday joys and worries of a fourteen-year-old, have all been blown away. I feel something that I've never felt before, and I don't understand what it is. I do know that I feel compelled to look at the picture for a long time and to look it up many more times. I also know that I don't want to tell anyone about this experience. It feels as if I have been filled with a new meaning that represents the whole world and at the same time also represents me. Suddenly, I'm no longer a schoolboy, a scout, a stamp collector, or a beginning pianist. What is the source of this unbounded rapture, this joy that forces everything else aside? Is it a presentiment of what I will feel some years from now when I touch a girl's knee for the first time, like that young man in Rodin's statue?

The Ring of the Nibelung

Budapest, 1940. "Siegmund, look at me! It is I whom you shall soon follow." Siegmund holds his sleeping bride, his sister

Sieglinde, on his lap and gazes with amazement at the majestic apparition of the Valkyrie, her visage hidden by her helmet. Brünnhilde announces that he must follow her to Valhalla, where only dead heroes are allowed, to bask in eternal fame, power, and heavenly joy in the company of Wotan and the other gods.

It is dark in the opera house, except for the light from a small, poorly shielded flashlight shining at the side of the highest row in the gallery, where there is only standing room. I am following the score, totally absorbed. I'm listening to the unforgettably unique voice of an erstwhile cantor from Stockholm, Set Svanholm, the director of the choir at Jacob's Church, a baritone converted to a great Wagnerian tenor. I hear his question, rising from the depths of fulfilled love, triumphant and yet forsaken, joyfully determined but with an incurable inner core of loneliness: "Attendeth her brother my bride and my sister? Shall there Siegmund Sieglinde find?" (*Begleitet den Bruder die bräutliche Schwester, umfängt Siegmund Sieglinde dort?*[2]) Brünnhilde answers No. Sieglinde must breathe earthly air for some time yet.

That voice again! The voice that we could wait for all night, lined up on the sidewalk in front of the opera house ticket office in Budapest—that voice that was never really appreciated in Stockholm, with its strange mixture of sadness and spirituality, the Heldentenor with the heart of a church musician. It begins deep down and grows slowly and inexorably in an unrelenting crescendo, like a basement fire that spreads upward to engulf a building. "Then greet for me Valhalla, greet for me Wotan, greet for me Wälse, and all the heroes. To them I'll follow thee not!" (*So grüsse mir Walhall, grüsse mir Wotan, grüsse mir Wälse und alle Helden! Zu ihnen folg' ich dir nicht.*)

Siegmund chooses one moment of love over eternal glorification, a conscious but fatal choice. Every other character in this most gigantic work of the operatic literature chooses gold over love: the Nibelung Alberich, the deformed son of the dark mist, who renounces love to be able to grab the Rheingold for himself; Wotan, highest of the gods, who promises away the love goddess Freia in return for the power and glory of Valhalla and thereby becomes entangled in insoluble conflicts; the giant Faffner, who kills his brother Fasolt to take the golden ring with which Wotan had hoped to consolidate his power even further. Is this the mythological sublimation of Wagner's own situation? We confront black and white, merciless dilemmas, and crises as in a Greek tragedy, but here it is part of a moral system without morals, an impossible pact that cannot be fulfilled, a world order without order, an eternal realm of the gods without eternity, an inexorable logic without logic. Love has the highest value, but it is persecuted by all and betrayed by all. Power and gold reign, but only to dig their own grave.

I am seduced by the music. Overcome by the powerful vision, I give in to its incomparable suggestiveness. I feel tossed helplessly back and forth on this ocean of feelings. Captivated by the story, I read myself deep into a world of motives, a pre-Freudian subconscious—but I understand nothing. Without any knowledge or premonition, I instinctively accept the notion that love is supreme, unreachable, damned, always relinquished, sold, abandoned, deceived; but the traitor, the deceiver, the man or god of power has nevertheless no chance to succeed. His path leads unerringly to Ragnarök, to be destroyed in flood and flame. [3]

Why is love both the eternal victor and the eternal vanquished? How can it simultaneously be the most desirable and the most forsaken thing?

Starry Eyes

Lucevan gli occhi suoi più che la stella
Her eyes surpassed the splendor of the stars
Virgil tells Dante of the visit of Beatrice, in *Inferno,* canto II [4]

Budapest, 1941. I follow Dante on his path toward pure, heavenly love, through Hell and Purgatory, sharing his terror and his relief. His metaphors speak deeply to me. Like the great Hungarian poet and Dante translator Mihály Babits, "I find splinters and shreds of my own soul in every corner." I am striving toward the meeting with Beatrice, but I find her face almost unbearably radiant. At the end of Purgatory, however, I myself am Dante, and in the company of Beatrice I am "pure and prepared to climb unto the stars" (*puro e disposto a salire alle stelle*).[5]

The Phoenix

Budapest, 1942. I am strolling on Margareta Island with one of my philosophical friends, discussing only serious things. Our emotional life doesn't come up at all. We don't waste our words, and we focus mainly on philosophy and literature. The discussion involves Spinoza, and we try, without any noticeable success, to understand a particularly difficult passage. Suddenly my friend looks at me and says, without any preamble or followup, "To go to a bordello is like murdering poetry."

I don't answer, and we continue our discussion of Spinoza.

I know that he's right, but he's also wrong. Poetry is like the Phoenix. It rises again after every conflagration.

The Tightly Bound Lovers, Number I

Was Dante a stranger to earthly love? Not at all. We don't have to read his biography to know that; we need not even read to the end of *Purgatory,* where Beatrice reproaches him for the years during which he forgot her. It is already evident in canto V of *Inferno,* in his conversation with the lovers who are damned for all eternity. How could Dante the pilgrim have been so touched otherwise; how could Dante the narrator have written about Paolo and Francesca as he did? He is already referring to this compassion, and to his compassion for all other condemned souls, at the start of the journey to Hell, as he prepares himself for the dual struggle against the hardships of the journey and against his own feelings:

Lo giorno se n'andava, e l'aer bruno
toglieva gli animai, che sono in terra,
dalla fatiche loro; ed io sol uno
m'apparecchiava a sostener la guerra
sì del cammino, e sì della pietate

(The day was now departing; the dark air
released the living beings of the earth
from work and weariness; and I myself
alone prepared to undergo the battle
both of the journeying and of the pity)

When he sees the lovers, eternally bound together and whirled about helplessly in an everlasting hurricane, Dante thinks about the glances and feelings that have brought them onto this perilous road. With the permission of Virgil, he asks them about this. He receives the most memorable response in all of the *Divine Comedy*: "There is no greater sorrow than thinking back upon a happy time in misery." (*Nessun maggior dolore, che ricordarsi del tempo felice nella miseria.*[6])

Francesca then tells about the book that she and Paolo read together on that fateful day, without suspecting any danger (*Senza alcun sospetto*). When they read about how love struck Lancelot, they were themselves overcome with the same feelings, which they could not resist. She talks about love as a higher power against which no individual has any defense: "Love, that releases no beloved from loving" (*Amor, che a nullo amato amar perdona*).

Paolo and Francesca haven't *chosen* their love, their entanglement, their sudden and violent death, their eternal damnation. It was determined by a higher power, which suddenly is called Amor. In the midst of this most Christian of all visions, the same name is used for both the heathen god of love and the beatific vision at the conclusion of *Paradise*—"the love that moves the sun and the other stars" (*l'Amor che muove il sole e l'altre stelle*).[7] Far from the unbearably radiant sublimation of Paradise, the reader joins Dante, the pilgrim, in the deep abyss with Paolo and Francesca, in the presence of this yearning, irresistible, delightful, and damning power. The book that by coincidence brought the lovers together is called their Galeotto, after a famous procurer.

Amor can obviously not be resisted, but it should have been avoided, according to this combined medieval and Renaissance logic, the logic of destiny in the Greek tragedy. The sinners, punished for all eternity, take their retribution as self-evident, but they show no sign of remorse. When Dante first sees them among the many couples who are blown by the eternal wind, and asks Virgil how he might come to talk to them, the Master advises that he should call them in the name of "that love which impels them" (*per quello amor che i mena*). They come immediately—and how!

Quali colombe, dal disio chiamate,
con l'ali alzate e ferme al dolce nido,
vengon per l'aer dal voler portate

(Even as doves when summoned by desire
borne forward by their will, move through the air
with wings uplifted, still, to their sweet nest)

Dante the pilgrim is shaken by Francesca's tale, possibly more than by anything else in the entire enormous realm of the Comedy. He describes it as follows:

Mentre che l'uno spirito questo disse,
l'altro piangeva sì, che di pietade
io venni men così com'io morisse.

(And while one spirit said these words to me,
the other wept, so that—because of pity—
I fainted, as if I had met my death.)

And then the dark words fall like a hammer blow, and with them falls Dante the pilgrim: *E caddi, come corpo morto cade* (And then I fell as a dead body falls).

The Tightly Bound Lovers, Number II

Stockholm, 1950. I am attending a lecture on parasitology given as part of the medical school curriculum. It starts as a somewhat dull matter-of-fact presentation of the second most widespread disease in the world: schistosomiasis, also called bilharzia. Only malaria claims more victims among the peoples of the so-called Third World. The lecturer is describing the glorious life cycle of the disease-producing worm, *Schistosoma mansonis*. It is both unbelievably horrible and wonderfully ingenious, the triumph of the worm and the degradation of the human. I see the picture of the male, with its mating furrow containing the much smaller and almost graceful female, the two tightly bound during their entire adult life span in incessant copulation in the human vascular system. My thoughts alternate between fascination with such perfect adaptation and nausea over the patient's terrible predicament, but then they leap suddenly to Paolo and Francesca.

I look around the lecture hall. My fellow students are taking notes in a matter-of-fact, diligent, essentially uninterested way, their eagerness focused only on our next exam. No one seems to have been seized by either delight or nausea. I'm the only one. Should I tell them about Francesca and Paolo during our break? I would only be laughed at.

But listen now and be amazed, *illustrissime lector!*

The worm's eggs, containing fully developed embryos, reach fresh water through human waste and infect snails. The larvae develop in the snails and are discharged after several metamorphoses. They look like small projectiles with a inverted V-shaped appendage at the tail. They are found most extensively in some lakes of China, the Philippines, Indonesia, Egypt, and many areas of Africa and South America. When a larva comes upon a human bather, it attaches to the skin. The tail drops off immediately and the body of the larva burrows into the skin. It enters the first small vein that it encounters. From there, it is carried through the blood to the lungs; then it proceeds to the heart and the general circulation. It can survive only if it reaches an artery that provides blood to the intestines and other abdominal organs. If it succeeds in lodging there, it can make its way further through the same circulatory system and penetrate even smaller blood vessels, until finally it comes to rest in the capillaries of the intestinal wall, whence blood plasma rich in sugars and other nutrients is transported from the digestive system to the liver. Here it feasts on all the delicacies provided by its environment. When it reaches maturity, the adult worm works its way back toward the bloodstream until it finds the walls of the small or the large intestine, depending on the species (the species most common in Japan prefers the small intestine; the African variety thrives only in the large intestine). The worm is now a sexually mature, graceful, sylphlike creature, between 10 and 25 millimeters in length, prepared for its love life. The male opens the long furrow along its "abdominal" side; the female inserts herself in the canal and remains there. They are not bound together by any infernal wind, but rather by a lifelong mating activity that results in perpetual egg production. There they live, in our vessels— yes, exactly so, *illustrissime lector*, it could be in you or me— constantly bathed by our warm blood, spreading their innumerable eggs throughout our blood vessels. The eggs escape the vessels as soon as they can, and are carried to the intestines, the bladder, the liver, the spleen, and sometimes even the brain.

But the human host, the unwitting and doomed guardian, knows nothing about the incessant love going on in his vascular system. He mobilizes his immune defense and inflammatory reactions against all these foreign bodies that appear suddenly here and there. This leads to an infernal, self-destructive battle conducted by the foreign-body reaction. That reaction can usually confine and eliminate dead particles that have succeeded in penetrating the sacred ground of the organism, but in this case the unending source of the foreign particles, the mating pair, survives in the best of health and keeps dispatching offspring, an unending flow of foreign particles, increasingly and without mercy. Finally, the emaciated host dies of his own allergic and other immune reactions. With him, the lovers also die. But before that, enough eggs have found their way through the intestinal tract to new waters and new snails.

The Severe

Mea Shearim, Jerusalem, 1985. Dark are the men of Mea Shearim. They have black hats, black hair, black beards, and long black coats. Even the small boys are dressed in black. They don't change their style of dress even during the most oppressive heat. The uncomprehending outside world cannot see the original significance of these garments any longer. This is how the Polish noblemen dressed at the time when Jews were granted civil rights and were allowed for the first time to dress in the same style. Their most recent descendants continue to demonstrate their freedom by a stubborn insistence on the same unsuitable style of dress. At best, they are indifferent to the modern secular state and its tolerance for all imaginable and unimaginable styles of dress. Some rather extreme minority groups in Israel distance themselves from the state and express their animosity and hatred through belligerent actions.

The walls around the main market in the district called "city of one hundred gates," Mea Shearim, are covered with anti-Zionist and anti-Israeli graffiti. Large posters urge the people to boycott the state's Independence Day celebrations—for those in Mea Shearim it is a day of mourning. On every street corner, from balconies, bulletin boards, and walls, multilingual signs urge any visiting women not to set foot in the district if they are not decently dressed. No jeans, no slacks, and nothing sleeveless. Even greater abominations are so unthinkable that they are not even mentioned. The religious fanatics from Mea Shearim have recently vandalized bus stations all over Jerusalem—some far from their own district—because the stations were adorned with what they considered to be enticing pictures of women in the ordinary commercial advertisements. They caused millions of dollars' worth of damage. The farsighted mayor, Teddy Kollek, who has done more for the cause of peaceful coexistence between Jews and Arabs and between peoples of different religious and political persuasions than any other Israeli politician, was beaten to the ground by a group of fanatics just as he was coming from a synagogue after a Friday evening prayer service. His "crime" was that he decided to build a public sports center in the secular part of the city where men and women could swim at the same time, dressed in very decent bathing suits. A prominent professor from the Hebrew University was stoned by a group of religious fanatics and suffered irreversible brain damage after they found him in the back seat of his car with his secretary, parked in a secluded spot.

And yet the Orthodox Jews are not hostile to sexuality. A man should be married and beget children. He should fulfill his marital obligations at least once a week, after a small prayer and a blessing especially designated for that purpose have been recited, provided that all the ritual purification injunctions have been followed. Mea Shearim and other religious areas swarm with children. There are young women, still in their twenties, already walking around with four or five children and almost always another one on the way. They wear long, simple gray or black dresses. Their heads are always covered, and many wear wigs instead of their natural hair.

My wife and I have been invited to an Orthodox wedding. She is just arriving in Israel and is taking one of the shared taxis that wait for six to eight passengers to assemble. The driver shows her a space in the front seat, where a dark figure in the characteristic outfit of the Chassidic Jews is already seated in the middle. He protests loudly and asks to be moved immediately to the back seat. It is inconceivable for him to sit next to a woman!

The wedding is simple, warm, relaxed, and intimate in a way that one would not expect with all this severity. The bridegroom and the bride are led in from different directions. The bridegroom is followed by his male friends; the bride arrives with her female attendants. The two meet again after not seeing each other for some time. Only a few decades ago, it would have been possible or even common that they had never seen each other before, or only at official family gatherings. When the music begins, the traditional wedding music of Chassidic Jews, the boys and the girls dance separately. Then, the young bride and groom are hoisted up in chairs, high above the boys' heads, and carried around with an enraptured, intoxicating joy that awakens Dionysian associations but is as far from the Greek culture as one can get in the Judeo-Christian world.

The Hospitable

Our work with the African childhood tumors during the 1970s led us into many unexpected collaborations. Because the shipments of tumor preparations and blood were expensive and involved a great deal of work for the clinic in Nairobi, we wanted to make sure that the material was used for as many research studies as possible. Thus, we passed the blood and tumors on to other laboratories in Sweden and elsewhere. Sometimes we functioned merely as a post office, sending out packages. At other times, we were drawn into interesting new collaborations. One particularly close collaboration developed with an American human geneticist, a Dr. F. He had developed a method that could tell us whether a tumor had originated from one single cell or from several cells. He compared certain genetic characteristics of tumor cells with those of normal cells from the same patient. Since only black donors had the genes which the method required, Dr. F received portions of all tumors and blood samples that came from Nairobi. For every blood sample from a tumor patient, a "control" blood sample was also sent from a different patient in Nairobi who was being treated for a non-malignant disease. The collaboration developed into a smoothly functioning routine.

One day, a very enthusiastic letter arrived from F. He had found a blood sample among the control material that contained a previously undescribed blood-group-like component. Now he wanted very much to obtain blood from the patient's closest relatives. It would be especially important to examine blood from any children that he might have.

We asked the secretary of the Swedish Radiation Clinic in Nairobi to look into the matter for us. It turned out that the patient was a Masai chief. He lived on the southern slope of Mt. Kilimanjaro and had three wives and nine children. Some eager young doctors at the Swedish clinic organized an safari to the remote village, and after several weeks we were able to send blood from the chief, his three wives, and all nine children to Dr. F.

After a couple of weeks, a certified letter arrived with a message written in red letters: "Highly confidential! To be opened only by Professor Klein personally."

Dr. F began by asking me to destroy his letter immediately after reading it. Under no circumstances was I to copy it, and he wanted to make absolutely sure that the information in the letter would never be passed on to Nairobi. So far, F's laboratory had tested blood from eight of the nine children and from all of the three wives. The Masai chief couldn't possibly have fathered any of the children. At the bottom of the letter was a postscript: "We have just completed the test on the blood of the ninth child. Paternity definitely excluded."

The little that I knew about the Masai people prompted me to make a copy of the letter. I sent it to our Swedish secretary in Nairobi, who had organized the expedition. I knew that she had a good sense of humor. She didn't fail me. At once she sat down at her typewriter and wrote the following letter:

Dear Dr. F!
The Masai are a people with highly developed moral and ethical rules which they follow strictly. But they're based on principles different from ours. When the Masai chief has invited a good friend for dinner, he takes the guest around the village and thrusts his spear into the ground in front of the hut of one of his wives. That signifies that he is inviting his friend to spend the night with the wife in the hut. If someone were to try to make his way into a wife's hut without an invitation from the chief, he would be killed immediately.

A postscript noted: "There is no reason to keep the results of the lab tests confidential."

The Puritanical

Stockholm, 1978. The state of Israel is celebrating its thirtieth anniversary and has sent Moshe Dayan as its representative to give a speech in Stockholm. I've been invited to an official dinner, and find myself placed next to the editor-in-chief of a major evening newspaper. We talk about the mood in Sweden for and against Israel, and I ask him if his newspaper receives anti-Semitic letters from its readers during periods of negative publicity about Israel. No, not at all. But there was indeed a torrent of anti-Semitic mail when the newspaper published excerpts from a new novel by Erica Jong.

I can't believe my ears, but my neighbor confirms that that I did hear correctly.

"Why is that?"

"The letters complained about 'Jewish immorality.'"

"Here? In Sweden, the Promised Land of pornography?"

"Absolutely, but there's a difference between pornography and pornography. Here in Sweden we have puritanical pornography."

"What in the world is that?"

"We've shed all our taboos. Everything that was previously forbidden is now allowed; everything that was previously concealed is now shown openly, but in a clean way. It's all anatomy and gynecology. But in the case of Erica Jong, it's mixed with shadowy and complicated emotions. The letter writers think that this leads to lascivious chaos."

I have been taught a new view of puritanism.

Nature's Lure, the Ultimate Illusion

I am visiting an institute for experimental brain research. The scientists have introduced electrodes into the brains of rats and placed them so that they can stimulate either single nerve cells or groups of nerve cells. They've found a "pleasure center" within an area of the brain called the hypothalamus. A rat with an electrode in this area is able to stimulate its own sensations of orgasm—which, contrary to what one would imagine, are formed in the brain. The rat can release the stimulating impulse by stepping on a little pedal. After the rat happens to touch the pedal accidentally as part of its ordinary activities, it soon discovers what sensations it can awaken. From that moment on, the rat cannot leave the pedal alone and doesn't bother with anything else. It stops eating and drinking and continues pressing on the pedal until it dies from exhaustion.

A Quiet Night in the Old Town

Stockholm, 1984. A warm and bright summer night. Walking toward Stortorget (the main market square in the Old Town), we are greeted by the unexpected sight of cars and people gathered near the great church called Storkyrkan. The police have set up riot fences between the church and a group of demonstrators standing outside and holding large signs. Two tall young men are standing on the church steps, embracing and kissing. From the church, we can hear the people singing "We Shall Overcome." It is the end of Gay Liberation Week. On the other side of the police cars, the demonstrators hoist their signs: "AIDS is God's punishment."

Great Genetics, or The Male's Little Ego

In the mid-1950s I received an urgent message from Professor Tracy Sonneborn in Indiana, summoning me to a "very important" meeting. The Nobel Prize winner H. J. Muller had come up with a new idea. Sonneborn and other geneticists felt that the idea should be killed and buried as soon as possible, but it deserved a first-class burial, worthy of its great creator. For that purpose, he was inviting me to an open symposium to include biologists, physicians, sociologists, philosophers, and even theologians.

H. J. Muller was one of the greatest figures of modern genetics. He had received the Nobel Prize for his discovery that radiation causes mutations. He was known as one of the most innovative, rigorous, and critical scientists in his field, the branch of genetics that uses the fruit fly as its experimental material. Many of the classical methods of this field had been worked out by him. However, there was another face to him. Whenever he perceived a great danger that might threaten the human gene pool, he took on his role of political agitator. This role had been very important shortly after the Second World War, when Muller had successfully warned the world about the genetic risks of radiation. At the beginning of the 1950s, this concept had become firmly established in our general consciousness. Muller's worry had now turned to another field. After all of the publicity about several of the worst abuses of genetics, including the mad ravages of the German "race biologists" and the literal eradication of scientific genetics in the Soviet Union, Muller had begun to assert his view that the human genetic material was in great danger and that certain corrective steps ought to be considered. It was his specific suggestion aimed at this goal that had aroused profound worry in his colleague Sonneborn.

We gather in a large auditorium at a small college in Ohio, a panel of about twenty people with approximately a thousand in the audience. Muller presents his thesis: The human gene pool is in danger of degenerating. The least intelligent, the most ruthless, the least considerate individuals are having the greatest numbers of children. People who are concerned about their children and their future usually plan carefully before they bring a very limited number of children into the world. This will eventually lead to a gradual deterioration of the gene pool. Muller's prescription for a cure of this trend is "germinal selection," and it includes the following features.

Approximately 30,000 women in the United States undergo artificial insemination every year. Each sperm donor is anonymous. As a first step, this ought to be changed. Instead, frozen sperm of deceased donors should be selected from a catalog. It is particularly important to preserve sperm from people who have made extraordinary achievements. Intelligence is the trait of highest priority for Muller. He leaves some room for distinguished characteristics of other sorts, but doesn't specify them any further. He also suggests that catalog selection for artificial insemination should be considered only as the first step. He is convinced that the successful use of the method will be "contagious," and he hopes that an ever-growing number of young couples will enroll voluntarily to have their first child from the catalog. University students will be the standard-bearers for this new movement.

During the coffee break, unconfirmed rumors circulate that such a movement has already started among the students at Muller's own university. Some say that young couples are even considering choosing Muller's own sperm for their first child.

My colleagues on the panel are very embarrassed. No one wants to buy Muller's proposal, but it is difficult to disregard Muller's great authority and his pioneering contributions in genetics. Some members of the panel pay a certain lip service to the idea, all the while searching for arguments against it without ever stating clearly what they think. This leads to some absurd arguments. One of the fathers of modern molecular biology, Salvador Luria, says the plan is dangerous because a dictator or a terrorist could easily poison a whole nation's sperm bank. Muller dismisses this argument as irrelevant; dictators and terrorists would use more drastic measures. Another panel member comes up with the argument that young couples who agree in principle with sperm selection are likely to quarrel over the choice of a donor from the catalog. Muller quickly brushes this aside—if they cannot agree on such an important issue, they're not suitable for each other and ought to get a divorce.

No one dares to touch upon the central question.

One of our most critical and admired master teachers in genetics has tried to convince us that the human genetic stock is in the process of rapid deterioration. He has suggested a series of measures which are, to say the least, problematic. But he hasn't shown any statistics to support his argument that the unfavorable characteristics he attributes to the most rapidly reproducing peoples or social groups are really genetic in origin rather than social or economic. He hasn't even attempted to discuss this issue.

At the end of the meeting, Muller asks to make some concluding remarks. He describes the young men who will voluntarily forgo fathering the first children of their marriages. Such young men are the most warm-hearted of all human beings. They have a higher regard for the interests of humanity as a whole than for their own little male ego. They have realized that the biological survival of the individuals through their children is an illusion, since the genes are diluted out very quickly and every trace of personal identity disappears after a few generations. The real altruists are those with a long-range vision of the future—they are our only hope.

I cannot stand it any more. If my older and more prominent colleagues won't do it, then I will have to speak up myself, despite being so young and green.

I ask to speak. I point out that the characteristics Muller attributes to those men who will voluntarily forgo siring their own children are certainly admirable, just as Muller has said. In fact, it is precisely these characteristics that we must promote. The solution therefore is obvious: we should choose *their* sperm first.

Why Sex?

Sexuality—one of our primary needs, the primal drive that encompasses the noblest and the basest, the most beautiful and the most sordid, the mainspring of poetry, art, literature, and music, the force that can create or destroy, much celebrated and much defiled—what is it really? What is the source of its success throughout the entire biological world? In Jerusalem, I ask Lea Ettinger, who has studied that question.

She reminds me that although sexuality has won out in the eukaryotic world, to which all higher animals and plants belong, it is not nearly as widespread or as important in the prokaryotic world, the world of bacteria. It is used, but relatively rarely, and then only as one of several alternative mechanisms.

One cannot give any simple or definitive answers. It was previously presumed that the recombination of genes between two individuals enhances the survival value of the species by promoting a greater diversity among individuals. Sexual reproduction would be superior to asexual mechanisms in this regard. Sexually reproducing species would have a great number of choices and could adapt more easily to new demands from the environment.

This explanation doesn't hold, says Lea. Even some of the simplest calculations of population genetics show that this idea must be incorrect. The reduction division that leads to the development of the male and female germ cells has a high cost in terms of "genetic currency." Every individual passes on only 50 percent of his or her genes to each child. Every person's genetic makeup is the finely polished product of many millions of years of environmental selection. The fact that half of this treasure is discarded each time a germ cell is produced indicates that sexual reproduction must bear some enormous advantages,[8] but these advantages are not at all evident. According to the mathematical analyses of John Maynard Smith,[9] the patterns of selection that occur commonly in nature cannot explain the advantage of sexual reproduction over asexual reproduction. Nor has it been possible to prove such an advantage experimentally. Microorganisms that can choose between sexual and asexual reproduction usually choose to multiply asexually when forced to adapt to new conditions.[10]

These analyses don't provide any evidence for the great success of sexuality—on the contrary. If they give a true picture, one would expect that sexuality would appear only now and then, always under very special ecological conditions, but not as the predominant mechanism for reproduction in all higher organisms. Some important factor is missing from the equation.

One has to search for this factor among the needs that all eukaryotes have in common, in both the animal and plant worlds. Lea has formulated a theory. Germ cells arise through a special reduction division that cuts the number of their chromosomes by half. Because of this highly controlled reduction, each germ cell receives only one of each kind of chromosome, instead of two as in all other body cells, and this reduction in the chromosome number is an absolute prerequisite for sexual reproduction. Fertilization restores the normal double number of chromosomes, with one full complement of chromosomes from the father and one from the mother. Lea's theory postulates that the great biological value of sexuality lies in this process of chromosome halving—that is, in the reduction division that occurs during the formation of the sperm and the egg. Conceivably, the reduction division may stabilize the gene material by setting definite limits to its dynamic flexibility; it may establish a strictly conservative order in a system that permits all imaginable (and many unimaginable) possibilities for radical structural change.

I find Lea's idea very attractive. It is based on the revolution in genetic concepts that began in the 1950s when Barbara McClintock made her original discoveries with the genetics of maize. Her work was not generally understood until two decades later. (She was finally awarded the Nobel Prize in 1984.) Until that time, it was thought that the chromosomes and their genes were very stable. Genes were thought to reproduce as rigid, highly constant, linearly arranged structures, generation after generation. It was believed that changes could occur only through mutations and, accordingly, were very rare. McClintock's experiment revealed that this stability was illusory, or rather that it was only a part of a more complex reality. Genes could change their structure and their linear order very dynamically. They could jump from one chromosome to another. Pieces of DNA originating from other chromosomes or from microorganisms in the external world, including certain viruses, could insert themselves near genes or even right into the middle of genes. This could increase, decrease, or qualitatively change the activity of the genes or their ability to be regulated. A given group of genes could suddenly reproduce out of synchrony with the rest of the genes and produce hundreds of copies of themselves through a process called "gene amplification."

The amount of the hereditary substance, DNA, is the same in all cells of the body. Most mammals contain approximately the same amount of DNA per cell. But in other classes of animals, such as the amphibians, different species can vary greatly in this respect. The salamander has almost 100 times more DNA per cell than the frog, but approximately the same number of functioning genes. This vast amount is due mainly to different amounts of "selfish DNA," or DNA without any function (see chapter 12). Much of the selfish DNA must have accumulated within the genetic material of the species through apparently quite accidental events, such as gene jumping or amplification.

Lea Ettinger thinks that there must be some mechanism that supervises the changes brought about by these movements and prevents undue chaos within the genetic equipment. Would it be possible to find the explanation for the incredible success of sexuality in such a control function? The process of reduction division is a good candidate. Before the chromosome number can be halved during maturation of germ cells, each chromosome has to seek its partner and line up in parallel with it. This pairing of the two corresponding chromosomes is a prerequisite for their correct separation by the "spindle apparatus" so that each daughter cell receives exactly one chromosome of each kind. Normally, each human cell contains 23 chromosomes from the father and 23 from the mother. An ordinary body cell doesn't distinguish between chromosomes from the two different parents, and the corresponding chromosomes of any given pair usually do not seek each other out during the normal process of cell division. They carry out their work in the cell nucleus independent of one another. But during reduction division, when new germ cells are about to be produced, the chromosome number has to be cut in half. Each germ cell receives one of each kind of chromosome. It doesn't matter if it comes from the mother or the father. Actually, the corresponding chromosomes from the parents (homologous chromosomes) are not retained in pure form; they exchange pieces by a process (called "crossing over") that occurs during the maturation process that leads to the reduction division. Therefore, the assortment doesn't divide up the chromosomes by their maternal or paternal origin, but there must be one chromosome of each of the 23 kinds—there cannot be zero of one kind and two of another. Mistakes can easily occur when two chromosomes stick together instead of separating normally. This may have catastrophic consequences. When one of the larger chromosomes is involved, the result is a deformed fetus that dies during the pregnancy. If one of the smallest chromosomes is affected, for instance No. 21 (through its inappropriate duplication in one of the germ cells), a fetus that has received three No. 21 chromosomes will have Down syndrome; a fetus with only one No. 21 chromosome becomes very severely deformed and is not viable. Only in the event that the asymmetry involves the sex chromosomes, X or Y, are the results less detrimental to normal physical and mental development, since its effect is often confined to the development of the sex organs.

How does each chromosome find and pair so exactly with its partner? This must be due to some kind of pattern recognition, the identification of certain landmarks on every chromosome. Lea has proposed that the so-called middle repetitive sequences in the DNA may be responsible. These sequences are distributed in different ways among the different chromosomes, but their arrangement is strongly conserved on each chromosome within a species—they are, in other words, identical in different individuals. The distribution of these sequences must be identical within the corresponding chromosomes of the father and the mother, to prevent faulty pairing between them. The consequence of the disarray that may result from major differences in the pattern can be best illustrated with hybrids between closely related species. The mule is a good example. A horse and a donkey are sufficiently close to each other to be able to mate and give birth to vigorous offspring that combine the advantages of the two parental species. But a mule cannot have any further offspring. Its sterility is due to differences between the chromosomes of the horse and the donkey. These differences are not great enough to affect the normal interaction between the chromosomes in the fertilized egg cell and during the whole process of development. The problem arises during the formation of the germ cells. The chromosomal "landmarks" are different, the reduction division becomes disordered, and the process of sorting the otherwise normally paired chromosomes into two equal halves does not function properly. Therefore, no functional egg cells or sperm can be formed.

Within a given species, these repetitive sequences may function as a built-in safety device against too-extensive "gene jumps" or other rearrangements of the DNA. These can change the chromosome structure to such an extent that it may interfere with correct chromosome pairing. The abnormal germ cell or its possible product, a chromosomally incomplete fetus, is eliminated.

The postulated importance of reduction division for "checking" and "correcting" possible damages to DNA is also supported by other facts. The DNA molecule has a built-in capacity for "proofreading." Some of its products can excise damaged or altered sequences and make a new correct copy of the text from the opposite, undamaged parallel strand (see chapter 12). It is known that such repair mechanisms, which exist in all cells, increase their activity during the maturation process leading to reduction division. It is also well documented that a significant proportion of the germ cells are eliminated during the process of maturation itself. This "weeding out" is so common that the process has been designated by name: atresia. Between 27 and 37 percent of all sperm cells and a much larger proportion of egg cells are destroyed before they are mature. This seems like a great waste, but it is not if it performs an important selective function. The extent of the atresia increases even further if the genetic material is damaged through irradiation, or in cases of inborn genetic defects.

Can this constant and repeated cleansing process explain why germ cells never age, in contrast to other cells of the body? Normal body cells have a limited life span, even in tissue culture. After a relatively small number of cell divisions (about fifty in the case of human connective-tissue cells), they have had time to accumulate so much damage or have undergone so many DNA rearrangements that they slowly and inexorably approach senescence and death. Germ cells remain forever young. This phenomenon, at first glance so paradoxical, becomes understandable if the germ cells are "tested" by a rigorous "quality control" as part of their maturation process.

The need to develop a control function that conserves the structural and functional integrity of the genetic material and thereby preserves it in a state of eternal youth must already have been essential for life when the first multicellular organism developed. The functional specialization of the body cells involves many risks of genetic wear and tear. The genetic material may be damaged by the external environment, or in the course of its own internal DNA dynamics. The problem is solved by separating the aging body cells from the eternally young germ cells.

Primitive single-cell organisms (prokaryotes) don't require a corresponding separation between germ cells and body cells, since every genetic modification is immediately subjected to selection by the external world. Only the most robust individuals survive. Higher organisms find themselves in a quite different situation. They make enormous investments in their offspring long before any possible deleterious effects of excessive genetic changes arise. The evolution of a selection process that eliminates deviant germ cells before they are fully mature must have been vitally important.

The need for quality control and eternal rejuvenation at the level of germ cells could also explain why all higher organisms develop from one single cell. It is worth noting that no multicellular organism has ever appeared during evolution whose reproduction was based on teamwork among several cells rather than on the division of one single fertilized egg cell. If reduction division serves as a checkpost for the integrity of the genetic material, it may explain not only the great success of sexuality and the eternal youth of the germ cells but also the rule that all multicellular organisms develop from a single cell in every generation.

The Eternal Spring

young men and women: *Eis aiona! Tui sum.*	In eternity, I am yours!
old men: *O res ridicula, immensa stultitia. Nihil durare potest tempore perpetuo cum bene, ...*	O what a ridiculous thing! What unbounded foolishness! Nothing endures the ravages of time. . . .
Catullus: *Mulier cupido quod dicit amanti in vento et rapido scribere oportet aqua.*	When a woman talks to a hungry, ravenous lover, her words should be written upon the wind and engraved in rapid waters.
young men and women: *Eis aiona, tui sum!*	In eternity, I am yours!

Carl Orff, *Catulli Carmina*[11]

In George Orwell's frightening utopia, every member of the party belongs to an "Anti-Sex League." Sexuality is suppressed and repudiated because Big Brother isn't able to control this private activity, one in which every individual can evade the surveillance of the state. When married party members want to have a child, they have to assume an air of self-sacrifice, weariness, and boredom and tell each other that "it's time to do our duty for the party."

But of course it doesn't happen this way, and it will never happen this way no matter how powerful Big Brother or his modern followers may be imagined to become in the kind of depersonalized, computerized society or technocratic system one might envision in the worst of all possible nightmares. Sexuality—nature's ingenious solution, the most wonderful and the deepest of all traps, mankind's common possession and yet each person's most private concern, this last refuge for many people today, their only opportunity to experience the illusions of eternity, this everyday experience that lacks any trace of mediocrity, this inexhaustible and eternal inspiration for all poets and artists—appears in the last analysis as the inexorable test of the individual before the tribunal of the species. Here, a person is reduced from a thinking and feeling being, with his own memories, sorrows, and joys and inherent uniqueness, to an unknowing germ cell, blindly propelled by an innate force, seeking its partner with whom it will pass through the needle's eye, the diamond-hard censorship of the chromosome-pairing game. Down turns the torch of Thanatos. The individual disappears, with all his conceited narcissism, while Eros leads the immortal, never-aging germ cell further on into the eternal spring.

14

Peter Noll's Awareness of Death and Wisdom of Life

Death, where is thy sting?
1 Cor. 15: 55

Wir haben keinen Grund Bewunderung und Liebe oder Hass dem Tod zu
zeigen den ein Maskenmund tragischer Klage wunderlich entstellt.
We have no grounds for showing admiration and love or hatred to
death, whom a mask's mouth of tragic lament grotesquely disfig-
ures.
Rainer Maria Rilke, "Death Experienced"[1]

Stockholm, autumn 1984. A thick German book appears in my
mailbox. The editors of the Stockholm publishing house
Brombergs Förlag want to know if I'd be willing to read it and
write a foreword for their edition.

I put the book aside and begin to go through the piles of
papers and letters awaiting my attention. Do I really have to
read such a big book, and in German no less? There's already
no end to the journals, review articles, and newly published
scientific books piling up on my desk.

I open the book without much thought and reach for my
dictating machine to draft a polite "no thank you." A rather
intriguing title, though, I must admit: *Diktate über Sterben und*
Tod (Dictations on Dying and Death). What sort of person
comes up with a title like that? I suppose I can spend at least a
few minutes on it.

Fifteen minutes later, everything has changed. The piles on
my desk have suddenly shrunk from Atlas-sized giants to Lilliputian figures.

Foreword to Peter Noll's Book[2]

Some years ago I visited a medium-sized but rapidly growing industrial city somewhere in Europe. I was with a committee charged to make recommendations on a proposed site for an international institution. We were shown around by the mayor's secretary. She was very anxious to tell us about everything the city was doing for its people. Here we're building a sports palace, here we're planning a cultural center, and over here we play tennis. This is our new concert hall, and here we've opened a new retirement home. The old people have a library in that wing of the building. Over there are the dining room and rooms for entertaining.

Some time later we passed a cemetery, and I couldn't help but imitate her: "Here is where we bury our dead."

We and they. We, still on this side, pretending that we're going to stay here for ever. They—those who have "left us"— have our valediction. *Vale,* farewell! Yes, certainly we send our farewells—go in peace , but *do* go! *We're* in a hurry, we want to continue building this unlimited future of ours.

One of the psychiatrists from the world's largest cancer hospital, in New York, a certain Dr. Ward, gave a talk in Sweden some time ago. He had been studying the adjustment of cancer patients to their disease and to their imminent death. He had come to the startling conclusion that cancer patients could be divided into two groups: Chinese and non-Chinese. The Chinese were prepared for death from an early age, whether they had been born in the United States or in China. They accepted the inevitable with great equanimity. On the other hand, the non-Chinese were not prepared. They either denied reality or reacted with great anguish. Some of them would put their faith in completely unrealistic and fraudulent treatments.

The Japanese frequently emphasize their Chinese cultural legacy. I therefore expected to find the same attitudes among them as among the Chinese. To my great surprise, I found that quite the opposite is true. Japanese cancer patients are not supposed to know the true diagnosis, and the white lie is generally accepted and practiced by Japanese cancer doctors.

It is justified by the danger that a patient will immediately commit suicide when he learns the true diagnosis. Physicians take it upon themselves to wage an energetic struggle to the very end, even where there is no rational hope left. It is no coincidence that Japan has the greatest number per capita of cancer researchers and physicians of any country in the world, and that Japanese surgeons perform superb and innovative cancer surgery. They also use a large variety of unconventional and uncritically tested methods and drugs, just in case they may help a little.

I once asked the grand old man of Japanese cancer research, Tomizo Yoshida, about the differences between Japanese and Chinese attitudes. This was several years before he himself developed incurable cancer. He agreed that there was a difference. After reflecting a while, he added that although the Japanese look up to the technological development of the West, they admire only the Chinese when it comes to existential questions and to philosophy of life. Why? What is it in the Chinese outlook that the Japanese admire so much, and strive to emulate but cannot reach? Yoshida was silent for a long while. I had already given up hope of getting an answer to my question. Then it came, suddenly and inscrutably: "Water birds fly over a lake. The lake's surface, my soul. The birds' shadows, the events."

Where does this serenity come from? Dr. Ward believed that the attitude of the Chinese cancer patients was based on their ancient family structure. One lives and dies as part of a large family in which all members have their individual privileges and obligations, depending on their sex and their order of birth—a system unchanged since time immemorial. It withstands all political pressures, trends, and fashions. The individual neither can nor wishes to break free, and neither illness nor any other catastrophe can change the basic structure. If a grandfather were to develop incurable cancer, it wouldn't occur to anyone to send him to a hospital to die. The grandchild will see him each morning before going to school. Yes, grandfather is about to die. But even if the death struggle takes a long time, there is nothing remarkable about that. The reality of death is softened by the continuity of the family. When the grandson becomes a grandfather, he will be able to accept his own cancer as equally natural.

Where do our Western attitudes come from? Has our culture impaired our ability to accept the unpleasant truth, the only one that applies to us all? Have we simply swept it under the rug? Must we lie about death just as we lie about life? Are there no other solutions?

Psychologists tell us that nonexistence is an emotionally unacceptable idea. An Eskimo tribe is said to have chosen the only logically consistent way out of this dilemma by completely denying the existence of death. They bury their dead right away and never utter their names again. They consider the dead never to have lived at all. Just the opposite is found in a certain Stone Age culture in New Guinea. Children eat their dead parents to allow them to continue to exist in their bodies. If they don't do that, it is taken as proof that they never loved their parents. Our Judeo-Christian culture has chosen to accept the idea of "life after death," a flexible concept with widely variable implications. These implications range from concrete and specific ideas to abstract allegories. All have the same purpose: to plant road signs where no roads exist, to install clocks where time has ceased, to provide the incomprehensible with a human face.

Is there no way out for modern man other than through the ostrich-like politics of denial, the delusion of superstition, or the mirage of mysticism? Does the pride of enlightened thought break like a fragile spider's web when the agony of death makes its entrance?

Peter Noll has chosen a different path. A universally recognized authority on criminal law, a successful academician, a famous author of textbooks, an expert frequently consulted by legislators, a man in excellent physical and mental condition, he learns at the age of 56 that he has bladder cancer. After considering all the pros and cons, he declines all forms of medical treatment, including the radical surgery still feasible for him, and devotes the remaining nine months of his life to his confrontation with death and dying. His limited time is filled with "a great deal of sorrow but also real bliss and, surprisingly enough, no despair whatsoever." He uses his time to dictate his book, but isn't sure whether he wants to have it published. Over and over he examines the purpose of the book. He doesn't want to seek comfort or extend his earthly life in other people's memories, an all-too-common but completely meaningless effort to keep going just a little while longer. He wishes to describe his experiences in a situation that sooner or later will affect us all. Most of all, he wants to convey his important insight that it is rewarding to deal with the problem of death and dying while one is still in the best of health. He even goes so far as to praise the dreaded cancer, since it still allows enough time to think about, and literally become acquainted with, death. Noll believes that we live better and more civilized lives if we live life as it is, with a constant awareness of its time limit. The duration of our limited time doesn't make any great difference, since we all fall short of eternity. We are all in the same boat. The only difference is that Noll knows approximately when he is to leave this world, whereas the rest of us live in blissful or unhappy uncertainty.

Only three months before his death, Noll writes that he does not yet know whether he has succeeded in imparting his insight concerning the importance of an awareness of death in enabling people to live a meaningful life. If he is convinced he has not succeeded, he plans to destroy the manuscript before his death. Readers who experience the catharsis of this book, as does the writer of these lines, will be grateful that he left his thoughts with us.

Noll's reasons for refusing medical treatment are very clear. He judges that his chances to resume a life of acceptable quality are very slim. He doesn't wish to lose his personal freedom by exposing himself to a long series of examinations and treatments that are incompatible with his concepts of life and death. Rilke, whose works were reportedly important to Noll, wrote in one of his incomparable prayer-poems: "Oh, God, give to each his own death" *(O Gott, gib einem Jeden seinen eignen Tod* [3]*).* Peter Noll has consciously chosen to receive this divine gift. He reacts strongly against the modern "mass-production" approach to dying. One dies on an assembly line, dying gets ever more banal, and individual death has been done away with like some unnecessary luxury. Death isn't any longer a subject for the living; it has been separated from life and relegated to the hospital and the nursing home. This is done in the hope that the survivors will be happier if they needn't be confronted constantly with death. In that regard we have been successful, to some extent, but at the same time we have lost something essential. We have merely developed a convenient device that belongs in the same category as washing machines and dishwashers.

With an extraordinary combination of dispassionate mental acuity, warm-hearted empathy, and the odd, melancholic, and sardonic sense of humor so often encountered in his book, Noll describes death as both the most universal and the most individual of all experiences. A very prominent person and a very ordinary person are equally dead when they're dead, and they are just as dead as a dead fly. Even so, dying is the most individual of all processes, since each of us must walk that path completely alone.

From his awareness of death, Noll looks out over the human landscape and perceives us, the living, in a new way. We all know that we must die, but since we don't know the exact time we ignore this fact and prefer to devote our time to our everyday problems. Our lives often seem meaningless. Those who are about to die never share this feeling. *Paradoxically enough, only the awareness of death can give life its true meaning.* This is Noll's most important message.

A man with limited time is aware of the chasm that suddenly opens between him and us. A change has taken place in a loosely defined and yet tacitly accepted basic principle. The rules of the game, which create a certain sense of togetherness, have been changed for one party, and others don't quite know how to behave. Conversations with a visitor become strained and suffer from mutual hypocrisy. Both are relieved when the visit is over. The healthy person is glad to be out of the sickroom, the dying person yearns for a little sleep.

Noll is aware that healthy people have a twofold problem when they visit him. In addition to the common need to repress their feelings, they face something that they sense as an "aesthetic dilemma." They are shocked, and would rather erase him from their consciousness. The scene doesn't conform to the accepted order of social conduct. If he were dead, everything would be solved. If he were in the hospital, like all other cancer patients, they could come bearing flowers. If he were to go home and later return to the hospital, they could once again come with more flowers, and so on, at ever-shorter intervals. Above all, they want to avoid seeing someone who is dying. But is the dying person obliged to respect these feelings? Does he have to do away with what is left of his body to make everything neat and tidy again? Should he start to regard himself as nothing more than waste? Noll poses the question unobtrusively, with great understanding and no bitterness, almost in passing, as if in parentheses.

Even the rituals of death are marked by a certain degree of embarrassment. Priests act like civil servants performing a task to help the living repress the reality of death and dying. In Noll's view, they should rather remind them constantly that any one of the assembled could be next, and that sooner or later they must all walk the same path. If they are ready, it may be much easier than they think.

Noll regards Seneca and Montaigne as his models. Since we don't know where and when death awaits us, *we* should await *it* instead. The conscious awareness of death enhances an individual's freedom. A person who has learned to confront it constantly cannot be submissive any longer. One who understands that the loss of life is not an evil need no longer suffer, as he might have before, when faced with wickedness. To live is to serve, and to die is to be set free. Why do these self-evident thoughts seem so extraordinary in modern times? Have we been subjected to a premature indoctrination—did we have rose-colored glasses put on us in the nursery? Similar thoughts have been expressed during all eras and in many different ways. Baudelaire put it this way:

C'est la Mort qui console, hélas! et qui fait vivre;
C'est le but de la vie, et c'est le seul espoir
Qui, comme un élixir, nous monte et nous enivre,
Et nous donne le coeur de marcher jusqu'au soir[4]

(Death? Death is our one comfort! is the bread whereby
We live, the wine that warms us when all hope is gone;
The very goal of life. That we shall one day die:
This is the thought which gives us the courage to go on)

The importance of an awareness of death for our ability to appreciate life is not limited to those who live in poverty or under oppression; it is equally relevant for us all. Noll notes how his own awareness of death has changed his concept of time, making it more valuable than anything else. One becomes much more selective about every undertaking and eliminates everything that is done merely for the sake of convention. One skips practical and useful but insignificant trivialities. Nothing is done for the sake of "being in style" if there's no true involvement. A relationship with a woman and conversation with good friends are all the more important. Your love grows for those who love you, and you are less concerned with those who don't. Many things become easier, some become more intensive. It becomes easier to tolerate things that have made you impatient before. You worry less about situations that used to make you anxious. You are more spontaneous and stronger in many situations where you may have previously been pliant and adaptable.

You watch television less and read more (fewer newspapers, many more books). You get more experienced at skipping quickly over pages where the author has simply written empty words and said nothing. The most important thing is to fill each moment with a meaningful and substantive content.

In the professional sphere, you try to minimize the time spent in meaningless meetings. Noll doesn't hide his contempt for most committees, which give a superficial impression of energy and resoluteness and conceal their indecisiveness behind forceful reports. In the imagination of Noll, Hell consists of an endless series of meetings. Satan is the chairman, and he sees to it that the committee doesn't decide anything that hasn't already been settled beforehand. One minor devil, specially selected for the task, prolongs every committee meeting with hopeless proposals and meaningless discussions of irrelevant issues or points that have already been decided.

How did Noll become such an eminent theoretical and practical jurist? What were his main driving forces? What is hidden behind such an impressive biography and collection of publications? The book provides some interesting clues.

As a young man, Noll wrote novels and dramatic plays. Like his close childhood friends Friedrich Dürrenmatt and Max Frisch, he prepared for a career as a writer but later became a lawyer. When the diagnosis of cancer was made, he thought first and foremost of the books that he still wanted to write. At the top of the list was a study of the importance of an awareness of death in an enlightened attitude to life. But he has also managed to say something about the second most important subject on his list: the eternal conflict between might and right that hides behind political demagoguery and refuses to yield to the unworldly theoretical analyses of legal philosophers. Privileges always have a tendency to accumulate in the hands of the powerful. In all societies, power is wielded by the mediocre, the established, and the uncreative. The machinery of their power is as clumsy and incapable of making simple and sensible decisions as their computers. They are merely floating around in a vast chaotic complexity. There isn't a single person who can assume responsibility for the systematic and colossal power structure that is, within the bounds of its own logically consistent frames of reference, anti-human self-destructive madness. The world will be destroyed by social planners, not by mad or alienated fanatics. Legal philosophers refuse to understand the true scope of power. It would be best to ignore their idealized reasoning. Only an analysis of the power play can help us, not the contemplation of abstract legal concepts.

It can be noted that Noll's deeply rooted and generalized conviction that "might makes right" is based on extensive experience with both democratic and non-democratic societies. His analysis has the same deadly acuity whether he deals with centralized power structures, revolutionary movements, or Lilliputian conflicts among the learned professors in Zurich. The fundamental play of power and adaptation follows the same ground rules. Nonconformists have only very meager chances to make their voices heard. Noll finds the only historical example to the contrary, the influence of a free spirit on the power structure, among the prophets in the Old Testament. Their effectiveness was due to the universal respect for their presumed divine inspiration. Today, they would be thrown into jail or into a mental institution.

Is Noll thinking of the widespread illusions of our times, particularly prevalent in the democratic societies, when he emphasizes the meaninglessness of confronting injustice with an idealistic view of justice? His logic is unrelentingly consistent. Injustice is the original condition of man. Justice must evolve laboriously through critical examination and reflection. Justice should rightfully be called non-injustice. It can exist only under conditions where injustice is resisted. This is often impossible, however, because the power structure can be confronted only by realistic analysis and with the establishment of counter-power structures. Legal philosophy never deals with anything like that. Instead, it devotes itself to meek and ineffective speculations on the concepts of justice and injustice.

I put the book down, my great pleasure from having read it mixed with sadness from not having had a chance to meet the author. But those feelings don't fully describe this reading experience, which has left me with a peculiar sensation. Have I been listening to a fascinating story? Yes, but that alone doesn't explain the feelings. Have I been touched by the author's fate, have I felt empathy with his situation? Yes, certainly, but that is still not the whole answer.

What could it be? Yes, now I know. Reading this book is like a long dialogue between the author and the reader, with the two on equal terms. "Never send to know for whom the bell tolls—it tolls for thee."[5]

I don't know if I will leave this world suddenly and unaware or slowly and gradually. But if I am going to lie ill for a long time before my departure, there are two books that I want to have by my bed. One is a volume of Rilke, including his poem on the gift of individual death. The other will be the book written by one who has received, understood, and implemented this gift: the book by Peter Noll.

15

The Atheist and the Holy City

The mist that came from the Mediterranean Sea blotted out the city
that Pilate so detested. . . . Jerusalem, the great city, vanished as
though it had never been. . . .
It had emptied its belly over Mount Golgotha, where the executioners had hurriedly despatched their victims. It had flowed over the
temple of Jerusalem, pouring down in smoky cascades from the
mound of the temple and invading the Lower City. It had rolled
through open windows and driven people indoors from the winding
streets. . . . But the flash passed in a moment and the temple was
plunged again into an abyss of darkness. Several times it loomed
through the murk to vanish again and each time its disappearance
was accompanied by a noise like the crack of doom.

Mikhail Bulgakov, *The Master and Margarita*[1]

I am standing on the terrace of my temporary quarters on
Mount Hadassah, near the large University Hospital. Down in
the valley I can see Ein Karem, the village of John the Baptist,
and Mary's well, where the Chassidic Jews come on one specific
day each year for water to bake their Passover bread. To the
right I can see the Russian Orthodox convent. The nuns live
nearby, but in a secluded world. I often see them walking up to
the bus station, two by two, their faces barely visible from inside
their black veils. They never speak with anyone, never look at
anyone. Some seem very old—I wonder how old they can be?
Ninety? One hundred? How long have they lived here, in their
very special Russian Orthodox Jerusalem?

Dark clouds hang menacingly over my head. Bulgakov has
never been here. But his vision of Jerusalem, written in Moscow,
is just as much present and yet just as absent as the Russian nuns.
The same is true for his Pilate, the lonesome procurator waiting

in the colonnade for the coming tempest, and also for most of us in the Judeo-Christian-Islamic world. Who among us doesn't find himself here, who among us doesn't long to come here, who among us doesn't yearn to get away from here?

"Next year in Jerusalem," Jews have been saying every year for many centuries. The Chassidic Jews of New York say it, those who could easily come here but don't and also those who do come. Even those who have already arrived in Jerusalem continue to express the same longing to come here.

Even the atheist lives in Jerusalem and always returns here. The word *Jerusalem* (*Yerusholayim,* in the ancient tongue) strikes a responsive chord that no other place name can.

What sort of chord could the atheist have?

"Golden Jerusalem, let me be your violin"—a song that pop singers and schoolchildren sing—is repeated in the popular tourist propaganda, but it never really seems to become vulgar. It seems to be strangely immune from trivialization.

The three great religions and their hundreds of often bitterly antagonistic sects, branches, and factions exist within a few yards of one another, but yet they have no channels of communication. Most of them exist as if in watertight compartments, separating like oil from water.

Only a few yards above the crowd in the bazaar, the swarthy Coptic monks sit like hermits in their garret atop the Church of the Holy Sepulchre. Their isolation is reminiscent of the caves in the desert. The ultra-orthodox Jewish *haradim* are also locked in, but in a very different inner world of their own, so dominated by their eternal dispute with their God. They reject and despise the godless state, the secularized world of sin.

The houses of the Moslems are fenced in and shut, turning their backs on the unfaithful. The Armenian quarter is closed to all strangers. Even those who belong are locked in or locked out after 6 o'clock every evening, and no one can enter or leave without permission of the archbishop.

As soon as the first evening star appears each Friday night, the Jewish prayers begin at the Wailing Wall. At the same time one can hear the recorded voices of the muezzins coming from awkward-looking loudspeakers atop the graceful minarets.

"Allah akhbar." The voices fly past one another. Words with identical meanings, wanting nothing to do with one another—they are light-years apart.

What can the atheist be seeking in Jerusalem?

I remember the words of a deceased colleague of mine who had grown up and worked in Jerusalem. He had survived the siege of the city and fought in its many wars. He was eternally at home and forever a stranger here, in this homeland for everybody and of no one. "Jerusalem is the most tragic of all cities," he used to say.

And what about all those old people who lie buried on the slopes of the Mount of Olives in the world's biggest open necropolis? Did they have time to put down roots before dying in the city of their eternal longing? How could they have done that during the Ottoman Empire as they awaited Death, their Liberator? Or during the Jordanian era, when their tombstones were vandalized, scattered about, or used to build houses, roads, or latrines? Are they now finally at home after the magnificent Israeli reconstruction, lying in orderly rows in the land of death, as close as possible to the places they had originally reserved for themselves to await the arrival of the Messiah on the day of the Last Judgment, in the most privileged site they could find, to follow him to the Holiest of Holies, through that golden gate temporarily sealed for the past thousand years? The graves offer no answer. They lie silent near Dominus Flevit, the church built on the site where the Lord wept, in a place now quite close to the neon lights of the Hotel Intercontinental. Is it really King David's city that they see in the valley below?

I am standing on my terrace watching dark clouds of migrating birds drift by, one after another. An ambulance helicopter on its way to the hospital disturbs their flight. I open myself to my own "Jerusalem mood"—a mixture of sadness and serenity, melancholy and confidence, defiance and resignation. I welcome it as an old friend. Every human being has his or her own "Jerusalem mood," a strictly private property as personal as an individual body scent that no dog could ever mistake, as unique as a fingerprint or as one's destiny. Never have I met anyone—

great or small, old or young, poor or rich, Jew, Arab, or Christian, mayor or street sweeper—who was indifferent to this place, a city that is no city at all, this tourist attraction where there are never any tourists, this political hornet's nest as far removed from the world of politics as Bulgakov's Pilatus from the local party committee in Moscow.

The fog rolls in directly above the Yad Vashem monument to the six million. A cold wind is blowing. I can still see the tall pillar that symbolizes the crematorium chimney at Auschwitz. I remember my grandmother and her sons. The thought of their last hours in the packed cattle car and of their last moments in the gas chambers remains indelibly tattooed somewhere in the back of my mind. I'm cold. I pull myself together and return to my warm apartment to open my comforting mail from Stockholm. I look forward to reading the latest news from my laboratory and of tumor biology. Jerusalem or no Jerusalem, that's enough for tonight! Back to reality, to the present, to the excitement of my science, to the latest reports from my colleagues, to my whole wonderful world.

But what are all these odd letters that I've received? Letters from strangers, stamped private letters, peculiar enclosures and brochures, periodicals from religious groups! An old lady from Skåne has sent me her mother's bible.

Ah yes, now I understand! The letters are from people who saw my TV interview with Per Ahlmark shortly before I left Stockholm. We talked a little about my past, and about cancer, and then we went on to the "great questions of life." I emphasized that I am an atheist. God is the greatest example of man's wishful thinking. We are alone in a cold and indifferent universe.

What is the meaning of life? Nothing whatsoever! Why should there be any meaning? What is the significance of meaning anyhow?

"But you are happy when you work? You work as if you were possessed. And you seem to be happy when you do it. How can that be?"

It is the same happiness that a dog must feel when running.

That's about how the whole interview went. But what are these letters trying to tell me? Do they come from people who are indignant or whose feelings I somehow have offended?

The letters vary a great deal in style and content, but they have one recurring theme, a kind of common denominator. It can be summarized as follows: You say that you are an atheist. But you work and behave as if you were religious. Therefore you *are* religious, whether you know it or not.

Several of the correspondents say that they feel sorry for me and want to help. All of them are quite positive and friendly. An old Nazi writes that he deeply regrets his youth. Only one of the letters is anonymous, but even that one is friendly.

I think about it. Is there anything to their claim about my "religiosity"? No, nothing at all. But anyhow, it's kind of them to write.

One of the letters is from a colleague, a devout Christian, a man whose attitude toward his work, his fellow man, and life itself I have always admired. He writes about as follows: "You are a scientist. You have to admit that it is not scientifically proven that God does *not* exist. Therefore, you should in all reason call yourself an agnostic, not an atheist."

I grab my pen and write: "I'm very sorry to disappoint you. I admire your work and what you are, but you should understand me quite clearly. I am not an agnostic. I am indeed an atheist. My attitude is not based on science but rather on faith, just as you have your faith. The absence of a creator, the nonexistence of God is my childhood faith, my adult belief, unshakable and holy. My faith is based on my experiences that have convinced me of the power of wishful thinking, our inability and aversion to accept hard facts, our desire to find extenuating circumstances."

It is Sabbath evening, and I am the guest of an Orthodox Jewish family, old friends of mine for many years. The father of the family, deceased for several years, was one of Israel's leading historians. Both sons—biologists—are sharp, intense, curious, and infinitely kind. They are deeply religious and always wear a kipa, the small knitted skullcap. The friendliness of their home reminds me of the Sabbath evenings at the home of my Orthodox grandmother, at another time, in another world and yet the same world. The same prayers are recited at the dinner table, the children sit with the same anticipation, and the father greets Sabbath, the bride, with the same delight.

I feel happy in this environment. I like my friends and admire much about their way of life. But, at the same time, it is difficult for me to understand their obviously deep belief in a personal God.

As usual, we talk about science, politics, and current events. But sooner or later we are always drawn into the inevitable and recurrent subject of our different views on religion, as if to an irresistible magnetic field. Now, for the first time, I realize with some astonishment that these Orthodox Jews are using the same arguments as those Christian letters from Sweden: You are religious, but you don't know it.

I deny it, almost desperately. What kind of God are you talking about? Where was your God while millions were being gassed at Auschwitz? I know that Orthodox rabbis have written long discourses on the subject. One of them said that God wanted to punish his chosen people for not obeying the call to Israel. Another equally venerable rabbi wrote that Auschwitz was God's punishment of the Jews for secularizing Jerusalem, profaning the sacred language, and establishing a godless state.

How can anyone believe in such nonsense? What kind of satanic God would concoct such punishments? Who would want a God like that? Isn't it high time to do away with him?

As usual, our long discussion leads nowhere. But we remain friends, and agree to continue to disagree. The brothers are adamant in their conviction that I am really a religious person after all, and they thoroughly believe that sooner or later I will *hozer b'tshuva* (return to the answer)— that is to say, find the right answers within the Jewish religion.

At one point late in the 1970s, I attended a meeting of the Swedish Scientific Advisory Council, convened to discuss the changing behavioral standards of Swedish society. There were talks by theologians, philosophers, historians, sociologists, and a few medical scientists. One of the speakers traced the relaxation of our standards to the inaugural speech of a newly appointed professor of philosophy in Uppsala around 1910, in which he "officially announced" the arrival of agnosticism. The belief in a personal God rapidly lost its grip on the intellectuals in the following years, but a puritanical morality, including the concepts of duty, hard work, responsibility, and order, continued to dominate the mood of the society for a couple of generations to come. The agnostic characters in the films of Ingmar Bergman, himself a minister's son, experience feelings of loyalty and treason, virtue and guilt, honor and humiliation within the same puritanical set of standards that existed during the times when parsons visited the homes of the people to test their knowledge of Luther's catechism. Their demons originate in the same Hell, and the choice between salvation and damnation is determined by the same fragile web of interactions. The theologians and philosophers were eloquent, but their presentations appeared strangely misplaced there in the large meeting room of the Swedish government. They maintained that puritanism and its work ethic will not survive in the long run. The message, as if from Cassandra herself, was clear: the father no longer transmits a strict work ethic to his son, and the old code of honor is being nibbled away.

The same theme was echoed with only minor variations. The principal speakers supported their conclusions with statistics, while the politicians resorted to short and disturbing anecdotes. When the meeting was almost over, one of the silent participants, a physicist, suddenly spoke up: "This is pure Kafka! We all know that all this is true, and we also know that there isn't a thing that we can do about it!"

Was the most important function of religion to promote ethical behavior? Will the Jewish ethical principle "Do not do unto others what you would not have them do unto you" or the Christian tenet "Love thy neighbor" vanish when all childhood faith has been relegated to the world of fables? Has the abolition of a personal God, one who watches over us from the cradle to the grave, opened the way for an increasingly selfish society? Are we risking a gradual disintegration of the social contract, a degeneration into a society that resembles the world of Idi

Amin in Uganda or, even worse, the non-society of the Ik tribe, whose ultimate moral degeneration is expressed as a contempt for everybody, with generalized malevolence and universal hatred?

It cannot be doubted that true faith can promote ethical behavior. But it is no guarantee. There is a man in Jerusalem who calls himself a rabbi but is shameless enough to agitate for racist laws along the lines of those proclaimed by the Nazi Party in Nuremberg—laws which he wants to be enacted by the same Jewish people who suffered most under similar laws. I also remember a Catholic priest, Father Kun, dressed in his cassock, helping members of the Arrow Cross to execute escaping Jewish boys in Budapest in 1944—hand grenades in his belt and a machine gun on his back.

I remember the Grand Inquisitor's Palace in Cartagena. After showing us the magnificent hall, the beautiful chapel, and the grandiose state rooms, the guide leads the unsuspecting visitors to see what he calls the "black pockmark" on the face of the Church: the gallows and the torture chamber. He watches the reactions on the faces of his audience with great delight while explaining how a huge wheel was used to crush living human bodies. After showing the thumb screw and a variety of other hellish instruments, he proceeds to the aquarelles from the same period. A priest of the Inquisition is shown questioning a suspected heretic while the executioners pour boiling acid through a large funnel into the ear of the accused. Other priests are rapt in prayer for the victim's soul, firmly convinced that they are performing a great humanitarian service. After all, the condemned heretic is on his way to salvation through the Inquisitor's torture and with the help of their prayers.

In the *Critique of Pure Reason*, Kant refutes a series of philosophical and theological arguments that had been used to prove the existence of God. He categorically rejects all metaphysical arguments concerning the immortality of the soul. But in the second book of the series, the *Critique of Practical Reason*, he accepts the existence of God on a non-logical basis. He derives his arguments from "practical reason" and, above all, from moral laws that require justice. According to these laws, happiness should correspond to virtue. This kind of justice can be administered only by a higher power, but obviously this has not taken place in this life. Kant therefore arrives at the amazing conclusion that God and an "afterlife" *must* exist so that the moral requirement for justice can be fulfilled. Man has a free will in his actions, since this is a condition for virtue and guilt.

Kant rejects the notion that ethical behavior could be motivated by utilitarian principles. According to him, morality has nothing to do with potential utility. It stems, instead, from reason, and is dictated by an imperative requirement to fulfill a duty. The most important tenet of his "categorical imperative" is that each person should act in such a way that the principles underlying his action should correspond to a universal moral law, acceptable to all. For example, it would be wrong to borrow money, since that action can never become a universal principle; if everyone were to borrow money, there would be no money left to lend. Virtue does not lie in the result of an action, but is exclusively dependent on the principle underlying the action.

Kant has often been regarded as the greatest modern philosopher. How could he accept the existence of God and the immortality of the soul on intuitive and irrational grounds and reject all arguments based on logical reason? In his *History of Western Philosophy*, Bertrand Russell points out that most philosophers after Plato considered it their self-evident task to examine the "proofs" of the immortality of the soul and the existence of God. They usually rejected the arguments of their predecessors. Saint Thomas contradicted the position of Saint Anselm and Kant repudiated Descartes, to mention two examples. They enunciated their own arguments instead. "In order to make their proofs seem valid," Russell writes, "they have had to falsify logic, to make mathematics mystical, and to pretend that their deep-seated prejudices were heaven-sent intuitions."[2]

How is it that Kant's "practical reason" accepted what he had already refuted by his "pure reason"? Did it depend on the conviction that only an ethical system based on religious faith can guarantee moral behavior, law, and order? Or could it have been—perish the thought—that Kant's settled academic life and his high position at the University of Königsberg may have prompted him to avoid, consciously or subconsciously, all unnecessary controversy, dispute, and practical difficulty?

An "atheistic rabbi," Sherwin Wine, is the leader of a Jewish congregation in Detroit. Wine declares that he has no interest at all in God, but that he identifies with the Jewish people and with certain aspects of Jewish tradition. Neither he nor his congregation recites any prayers, but they gather in the synagogue every Saturday. Contrary to Jewish tradition, they do not regard the Sabbath as holy, but rather as a day for the family and remembering the extended family of all humanistic Jews.

Rabbi Wine has met with a great deal of opposition. He has been forbidden to speak publicly and threatened with excommunication, but this could not stop him. His congregation now has more than 800 members, and similar Jewish groups have sprung up in other parts of North America. A 10,000-member society for humanistic Judaism has been founded, with Rabbi Wine as its chairman.

Humanism is Rabbi Wine's only true creed. He wishes to remain within Judaism, but only on the basis of its humanistic aspects. The conviction that truth can best be discovered with the help of reason is the most important of his tenets. Common sense dictates that a meaningful human life must be built on basic human needs rather than on obedience to divine bidding. The human being should be in the center, not God or religion. Here, Wine is deliberately breaking with traditional Judaism, whose authoritarian ethic holds that right is right because God has said so. His congregation does not accept the Ten Commandments without reservation. The first commandment, "I am your God": no, thank you very much. The second, which expresses God's jealousy: no, absolutely not. Thou shalt not kill, thou shalt not steal: yes, thank you.

Humanistic Judaism was born after the Holocaust. One of the most important sources of the movement was the book *After Auschwitz*, written by Rabbi Richard Rubinstein. The fact that after Auschwitz most people had to learn to live without God has made the religious laws more important than ever before, in his opinion. Wine goes even farther. He reacts to all talk about God with a shrug of his shoulders and adds: "A God whose existence has to be saved through mental gymnastics or anti-intellectual leaps of faith is not worth having." He has little or no interest in theology; he prefers psychology, anthropology, philosophy, and history. In contrast to most ordinary rabbis, he has no interest whatsoever in telling people what they should think, how they should vote, whom they should marry.

Lev Bearfield, a reporter for the *Jerusalem Post*, asked Wine why so many young Jews choose to return to Orthodox Judaism. Wine attributes this to fashions that appeal to people of a certain temperament. In the United States, it is often based on a kind of nostalgia. There are always people who do not want to take responsibility for their own lives but prefer instead to be instructed how to behave correctly. Wine sympathizes with them to some extent, but says: "If a person chooses safety and security at the expense of his dignity and his individuality, don't ask me to respect him."[3] He has never seen a humanistic Jew "go back" to religion. Khomeini could not solve the problems in Iran, nor can Falwell do so in the United States. Fundamentalistic Orthodox Judaism can only damage Israel. The religious fanatics threaten the beautiful concept that led to the establishment of the state of Israel. Their activities are a blow against the recognized unifying identity; they represent "a frightening and terrible irony." Wine therefore has taken the personally difficult step of establishing a center in Jerusalem, a move based on the conviction that "it isn't enough merely to be anti-religious." One must build a positive force, humanism, that can counteract the religious influence in the schools, the army, and the government.

In his already classic book *The Selfish Gene*,[4] Richard Dawkins emphasizes the inherent selfishness of the genes, their primal tendency to work for their own survival and replication. All organisms, plants as well as animals, might be considered as "survival machines," directed by the selfish genes. Organisms are discarded like empty cartridges after fulfilling their role as a temporary clothing for the genes. Dawkins cites many colorful examples of gene variations and their interplay with selection: harmless butterflies that disguise themselves in the clothing of their poisonous cousins to fool hungry birds, cuckoo chicks that not only behave like parasites in the alien nest but even annihilate its rightful inhabitants, ant colonies that kidnap slaves rather than work themselves—all this constitutes a part of the panorama. The apparently altruistic, self-sacrificing "kamikaze pilots" of the bee colony are driven by sophisticated but basically selfish genes. At first glance, the same holds true for the gentle gazelles that gather in large herds, apparently to protect individuals against predators.

After having been amused and even a bit horrified by Dawkins' many detailed and entertaining examples, one could summarize the "categorical imperative" that he attributes to the genes in the following way: "Body, follow a strategy that increases my chances for survival, at any cost. Pursue other creatures or flee from your pursuers; be loyal and faithful to others or cheat and falsify ruthlessly; protect and defend others or kill unmercifully; join alliances or break even the most sacred defense pacts; build or destroy; live in a society driven by mutual give-and-take or kill your parents and eat your own children—it doesn't matter, as long as your actions benefit my successful reproduction. After you have led me to develop new survival machines, I will discard you without hesitation. I am in constant jeopardy of losing out in the race against other selfish genes that have provided their survival machines with more effective competitive devices."

Dawkins succeeded in translating the relatively inaccessible mathematical language of the population geneticist John Maynard Smith into more easily understandable terms. "Evolutionarily stable strategy" (ESS) is one of the most important concepts. The word *strategy* means a programmed behavioral pattern, one example of which can be described as follows: "Attack your opponent. If he flees, pursue him. If he retaliates, run away."[5] The animal, the survival machine, follows this strategy instinctively. Our process of expressing this concept in words should not be confused with an awareness that doesn't exist.

ESS can be defined as a strategy whose evolutionary value cannot be improved by alternative strategies—if the majority of the population follows the strategy. One can also say that the optimum strategy for the individual depends upon the behavior of the majority.

It would be beautiful, suggests Dawkins, if human society were entirely composed of "doves" without "hawks," as in the dream of the Garden of Eden. (Dawkins uses these bird names according to common usage, although he points out that they do not correspond to real life. Doves are in fact very aggressive animals!) A dove society cannot survive in the long run if there are hawks in the population or if "hawk genes" appear through mutation. Pure "dove behavior" is therefore not a stable strategy. Simple mathematical simulations have shown that stability is attained only if a certain equilibrium exists between hawks and doves. In somewhat more complicated models, the population is supplemented with additional "types"—for example, the "retaliator," who plays the dove in the beginning but retaliates if attacked; the "bully," who behaves like a hawk until attacked but then immediately runs away; and the "prober-retaliator," who behaves like a hawk against an opponent who does not fight back (like a dove in the face of an menacing opponent) but who retaliates when attacked. A stable ESS situation is established at a certain optimal proportion of retaliators and prober-retaliators. Neither hawks, nor doves, nor bullies alone are stable as pure populations.

Dawkins says that he would have preferred to describe a different and better world, or at least a world more consistent with our expectations and our hopes. But he can only describe the world that is.

Does he leave us without any hope? Does he march out with a derisive laugh while we sink into the pitch-black pessimism that is so easily evoked by the reading of his wonderful book? No, not entirely. It is true that he devotes only the last page of the book to the future of man, but he does so on quite a promising tone. He ends with a "note of qualified hope." He points out that mankind has some unique characteristics, among which is conscious foresight. In contrast, "Selfish genes have no foresight. They are unconscious, blind replicators." Their most important property, that of self-replication, must inevitably lead to behavior that can only be described as selfish. No single gene can forgo its selfish behavior, even temporarily, even though a different behavior would bring it some advantage in the long run. A "conspiracy of doves" that would lead to a pure dove society would certainly be better for every single individual than the "mixed" society that provides a stable strategy. But as long as the short-term selfishness of genes dominates, a pure dove society would be unstable and would soon be destroyed through the process of natural selection.

To Dawkins it seems possible, although not certain, that mankind has the capacity to place a high priority on true altruism instead of mere self-interest. If this were true, it would be yet another of man's unique qualities. We may hope that this is so, but Dawkins is not prepared to argue one way or another. To be on the safe side, he prefers to keep to the dark side and "assume that individual man is fundamentally selfish." But even if that is so, our conscious foresight, our capacity "to simulate the future in imagination" in some detail, protects us from the worst excesses of our blind genes. We have mental equipment that can, at least in principle, "foster our long-term selfish interests rather than merely our short-term selfish interests." If we all realize that a conspiracy of doves is to our benefit in the long term, we can "sit down together to discuss ways of making the conspiracy work." This theoretically gives us the potential to defy our selfish genes and even their selfishly conceived conceptual products (which are transferred much more quickly than the genes themselves, through cultural influence). We can seriously discuss ways to promote something that has no traditional place in nature and which in reality has never existed in the entire history of the world: "pure, disinterested altruism." Despite the fact that we are survival machines for our selfish genes, we may have "the power to turn against our creators." Dawkins concludes with the words: "We alone on earth can rebel against the tyranny of the selfish replicators."

I am standing in front of the mayor of Jerusalem, Teddy Kollek, in his office—which is still in the same corner house in the Mamilla quarter of the city that he occupied before the wall that divided Jerusalem was torn down, in 1967. There are two large photographs hanging on the wall: the Jerusalem wall at the moment of its demolition, and part of the Berlin Wall (still standing at the time). Words are superfluous in the face of these eloquent images.

I remember my first visit to this part of the city, in 1962. I was standing on a roof above the temporary medical school's cramped and shabby laboratories. With the anatomy auditorium directly under my feet, I could look right down the rifle barrels of the Jordanian guards on top of the old city wall, hardly a hundred yards away. The guards were lying behind sandbags, with their rifles sticking out. Below me were the Israeli students in the lecture hall, listening to the eternal saga of the human body. Now and then the guards shot a few salvos at the high windows of the lecture hall, but not too often. Only when they felt the urge. No self-respecting Israeli professor would turn his head or acknowledge hearing the shots at all; he would lose the respect of the students forever.

I could see the wide street behind the wall on the Jordanian side. It started blindly and unnaturally between the city wall and the monastery of Notre Dame and led directly toward the Arab bazaar quarter. It seemed suddenly as if I had a glimpse of the back side of the moon, a forbidding, strange, inaccessible world. Today, my biggest problem is where I should park my car when I drive down that very same street to go shopping in the Arab market.

Teddy Kollek shows me the new Arab health center, which the Jerusalem Foundation has built, at his initiative, to serve the Arab population of East Jerusalem. A new and surprisingly beautiful structure built in Arabic style, it blends naturally into the landscape at the foot of Mount Scopus while retaining its own peaceful individuality. It is run by Arab doctors and nurses, accepts only Arab patients, and serves only Arab food. All signs and messages are in Arabic. Only certain laboratory positions are staffed by Jewish-Israeli personnel.

We go past the outpatient maternity clinic. Young pregnant women are sitting in the waiting room, some of them with their husbands. One of the doctors explains that, at the beginning, the husbands always come along. After a few visits, when they are convinced that everything is proper, they dare to let their wives come alone.

A strange calm and a certain quiet serenity prevail in the corridors and waiting rooms of the hospital. Even the pediatric department seems to be quieter than in the highly intense Jewish hospitals, with their loud, spontaneous expressions of joy, sorrow, pain, and anger. The Jewish hospitals also admit Arab patients—I see them often at the Hadassah Hospital, where I work—but they seem to be quite lost there. Here they feel at home. I tell Teddy Kollek enthusiastically that this certainly is a beautiful example of collaboration. Kollek looks at me sternly and says emphatically: "It's not a question of collaboration. We're hoping to achieve a certain degree of practical neighborly coexistence."

I begin to understand his pragmatic attitude. His goal is to get the many opposing groups—Arabs and Jews, religious and non-religious Jews, the Ashkenazim of the West and the Sephardim of the East, and the innumerable other religious denominations—to live side by side, in peace. He meets with the approval of the liberals but is condemned by the fanatics on all sides. The Palestinian Liberation Organization has urged the Arab population to boycott the health center, but without success. Kollek is also opposed by the Jewish religious extremists who think that they themselves know best how their one true God wishes it to be.

One can't help concluding that only a pragmatic politician like Teddy Kollek has a chance to bring about peace in this city which has been so sorely tested for thousands of years. But what would happen then? Would Jerusalem become a city like all other cities? It is an interesting question, but unfortunately still completely theoretical. The name Yerusholayim means "the city of peace." The translation remains as ironic as ever. The injunction at the beginning of Handel's *Messiah*, "Speak ye comfortably to Jerusalem; and cry unto her that her warfare is accomplished," is as current as ever. Will peace remain an unattainable dream in a city that is holy to so many different religions, just because it *is* holy? Is the unworthy struggle among Christ's many servants in the Church of the Holy Sepulchre, the unending squabble for every square inch, every entrance, and every doorpost, the eternal hallmark of this city?

It is evening again. The light over the Judean desert fades slowly. In the distance the light blue speck of the Dead Sea transforms its improbable presence into a more natural nothingness. The voice of the muezzin calls out from all the minarets. Dark clouds hang over Yad Vashem. Grandmother, are you there? Have you vanished in vain with the smoke of the crematorium chimney? No, it's not so. It is your death and the death of the other six million that awakened this ancient land from its 2000-year sleep and that has united divided Jerusalem into a city where good neighborly harmony is at least theoretically possible. It was your coming death that permitted Leo Szilard and the other physicists to lead us bravely, wisely, and foolishly into the atomic age. It was Auschwitz and Hiroshima that tolled the great doomsday bell for us all as our century approached middle age. Will this bell finally awaken our collective foresight, our basically selfish altruism, our only hope? The millennium is coming to a close. It remains to be seen.

Notes

Preface

1. Peter Noll, *In the Face of Death*, tr. H. Noll (Viking, 1989). (See also chapter 14 below.)

Chapter 1

1. Albert Levan and J. H. Tjio, "The human chromosome number," *Hereditas* 42: 1, 1956.

Chapter 2

1. *Leo Szilard: His Version of the Facts. Selected recollections and Correspondence*, ed. S. R. Weart and G. Weiss-Szilard (MIT Press, 1978).

Chapter 3

1. Ibid.

Chapter 4

1. Ibid.

2. Jacques Monod often commented later that Leo Szilard was one of the most important advisors that he and Jacob had in relation to the experiments on gene regulation that later earned them the Nobel Prize.

Chapter 5

1. In German, *Spiegelmann* means "mirror man."

2. Arthur Koestler, *The Thirteenth Tribe* (Hutchinson Press, London, 1976).

3. Mihaly Csikszentmihalyi, "Reflections on enjoyment," *Perspectives in Biology and Medicine* 28: 489–497, 1985

Chapter 6

1. Benno Müller-Hill, *Murderous Science*, tr. G. R. Fraser (Oxford University Press, 1988).

2. Benno Müller-Hill, "Kollege Mengele—nicht Bruder Eichmann," *Magazin*, June 15, 1985, p. 11.

3. J. Hallervorden, *Alg. Z. Psychiatrie* 124: 289, 1949.

4. R. Höss, *Kommandant in Auschwitz* (M. Borszat, Munich, 1948). In English: *Commandant of Auschwitz* (Pan Books, London, 1961).

5. Hugh Trevor-Roper, "Seas of unreason," *Nature* 313: 407, 1985.

6. Benno Müller-Hill, *Die Philosophen und das Lebendige* (Campus-Verlag, Frankfurt, 1981).

7. *Die Wissenschaften und der Holocaust* (Herbert A. Strauss).

Chapter 7

1. *Kalevala*, tr. W. F. Kirby (J. M. Dent, New York).

2. Marcel Proust, *Remembrances of Things Past*, tr. C. K. Scott and T. Kilmartin (Vintage, 1981).

Chapter 10

1. S. E. Luria, J. E. Darnell, D. Baltimore, and A. Campbell, *General Virology*, third edition (Wiley, 1978), p. 2.

Chapter 11

1. Ibid.

Chapter 12

1. For discussions of "selfish DNA," see W. F. Doolittle and C. Sapienza, "Selfish genes: The phenotype, paradigm and genome evolution," *Nature* 284: 601, 1980; L. E. Orgel and F. H. C. Crick, "Selfish DNA: The ultimate parasite," *Nature* 184: 604, 1980.

2. Arthur Schopenhauer, *The World as Will and Idea*, tr. R. B. Haldane and J. Kemp (Routledge & Kegan Paul, London, 1964), p. 191.

3. Ibid., p. 213.

4. François Jacob, "Evolution and tinkering," *Science* 196: 1161, 1977.

5. G. G. Simpson, "How many species?" *Evolution* 6: 342, 1952.

6. Schopenhauer, *World as Will and Idea*, p. 143.

7. Orgel and Crick, "Selfish DNA: The ultimate parasite." *Nature* 184: 604, 1980.

8. Schopenhauer, *World as Will and Idea*, p. 230

9. Ibid., p. 230.

Chapter 13

1. "Eternel Printemps" is generally known as "The Kiss" in English-speaking countries.

2. Translated from the German in *The Opera Libretto Library* (Avnel Books, New York, 1980).

3. In Nordic mythology, Ragnarök refers to the final fate and destruction—the twilight of the gods and the ruin of the world. It corresponds to the Götterdämmerung of Wagnerian legend. The word is derived from the Edda *ragna rök,* "the fate of the powers."

4. *The Divine Comedy* of Dante Alighieri, tr. A. Mandelbaum (Bantam, 1988). *Inferno,* canto II.

5. Ibid., *Purgatory,* canto XXXIII.

6. Ibid., *Inferno,* canto V.

7. Ibid., *Paradise,* canto XXXIII.

8. G. C. Williams, *Sex and Evolution* (Princeton University Press, 1975).

9. J. Maynard Smith, *The Evolution of Sex* (Cambridge University Press, 1978).

10. G. Bell, *The Masterpiece of Nature: The Evolution and Genetics of Sexuality* (University of California Press, 1982).

11. *The Poems of Catullus,* tr. H. Gregory (Grove, 1965).

Chapter 14

1. Rainer Maria Rilke, "Death Experienced." from *New Poems, 1907,* tr. E. Snow (North Point Press, San Francisco, 1984).

2. U. S. edition: *In the Face of Death* (Viking, 1989). The book was published by Pendo Verlag of Zurich in 1984 and by Brombergs Förlag of Stockholm in 1985. Klein's foreword does not appear in the Viking edition; it is presented here in translated form.—*Editor*

3. Untitled poem, in *Stundenbuch.*

4. Charles Baudelaire, "The Death of the Poor," from *Flowers of Evil,* ed. G. Dillon and E. St. Vincent Millay (Harper, 1936).

5. John Donne, *Devotions upon Emergent Occasions,* XVII.

Chapter 15

1. Mikhail Bulgakov, *The Master and Margarita,* tr. M. Glenny (Collins and Harvill, London, 1967).

2. Bertrand Russell, *History of Western Civilization* (Allen & Unwin, London, 1961), p. 789.

3. Lev Bearfield, "Saturday the Rabbi has nothing to worship," *Jerusalem Post,* July 27, 1985.

4. Richard Dawkins, *The Selfish Gene* (Oxford University Press, 1976).

5. Ibid., p. 74.

Index